Scripting Intelligence

Web 3.0 Information Gathering and Processing

■ ■ ■

Mark Watson

Apress®

Scripting Intelligence: Web 3.0 Information Gathering and Processing

Copyright © 2009 by Mark Watson

ISBN-13 (pbk): 978-1-4302-2351-1

ISBN-13 (electronic): 978-1-4302-2352-8

Printed and bound in the United States of America 9 8 7 6 5 4 3 2 1

Trademarked names may appear in this book. Rather than use a trademark symbol with every occurrence of a trademarked name, we use the names only in an editorial fashion and to the benefit of the trademark owner, with no intention of infringement of the trademark.

Java and all Java-based marks are trademarks or registered trademarks of Sun Microsystems, Inc., in the United States and other countries.

Apress, Inc., is not affiliated with Sun Microsystems, Inc., and this book was written without endorsement from Sun Microsystems, Inc.

Lead Editor: Michelle Lowman
Technical Reviewer: Peter Szinek
Editorial Board: Clay Andres, Steve Anglin, Mark Beckner, Ewan Buckingham, Tony Campbell,
    Gary Cornell, Jonathan Gennick, Michelle Lowman, Matthew Moodie, Jeffrey Pepper,
    Frank Pohlmann, Ben Renow-Clarke, Dominic Shakeshaft, Matt Wade, Tom Welsh
Project Manager: Beth Christmas
Copy Editor: Nina Goldschlager Perry
Associate Production Director: Kari Brooks-Copony
Production Editor: Ellie Fountain
Compositor: Dina Quan
Proofreader: Liz Welch
Indexer: BIM Indexing & Proofreading Services
Artist: Kinetic Publishing Services, LLC
Cover Designer: Kurt Krames
Manufacturing Director: Tom Debolski

Distributed to the book trade worldwide by Springer-Verlag New York, Inc., 233 Spring Street, 6th Floor, New York, NY 10013. Phone 1-800-SPRINGER, fax 201-348-4505, e-mail orders-ny@springer-sbm.com, or visit http://www.springeronline.com.

For information on translations, please contact Apress directly at 2855 Telegraph Avenue, Suite 600, Berkeley, CA 94705. Phone 510-549-5930, fax 510-549-5939, e-mail info@apress.com, or visit http://www.apress.com.

Apress and friends of ED books may be purchased in bulk for academic, corporate, or promotional use. eBook versions and licenses are also available for most titles. For more information, reference our Special Bulk Sales–eBook Licensing web page at http://www.apress.com/info/bulksales.

The source code for this book is available to readers at http://www.apress.com.

To Carol, Julie, David, Josh, Calvin, and Emily

Contents at a Glance

PART 1 ■■■ Text Processing

PART 2 ■■■ The Semantic Web

PART 3 ■■■ Information Gathering and Storage

PART 4 ■ ■ ■ Information Publishing

PART 5 ■ ■ ■ Appendixes

Contents

PART 1 ■■■ Text Processing

PART 2 ■■■ The Semantic Web

PART 3 ■■■ Information Gathering and Storage

About the Author

**MARK WATSON** is the author of 15 books on artificial intelligence (AI), software agents, Java, Common Lisp, Scheme, Linux, and user interfaces. He wrote the free chess program distributed with the original Apple II computer, built the world's first commercial Go playing program, and developed commercial products for the original Macintosh and for Windows 1.0. He was an architect and a lead developer for the worldwide-distributed Nuclear Monitoring Research and Development (NMRD) project and for a distributed expert system designed to detect telephone credit-card fraud. He has worked on the AI for Nintendo video games and was technical lead for a Virtual Reality system for Disney. He currently works on text- and data-mining projects, and develops web applications using Ruby on Rails and server-side Java.

Mark enjoys hiking and cooking, as well as playing guitar, didgeridoo, and the American Indian flute.

About the Technical Reviewer

■**PETER SZINEK** is a freelance software developer. He left his Java job and academic career a few years ago to hack on everything Ruby- and Rails-related, and never looked back. He is the author of Ruby's most popular web-scraping framework, scRUBYt! (http://scrubyt.org), which is featured in this book. After founding two startups, he started his own consultancy called HexAgile (http://hexagile.com), which offers Ruby, Rails, JavaScript, and web-scraping services. In addition to coding, he also enjoys writing—namely, blogging at http://www.rubyrailways.com, working on the "AJAX on Rails" guide for the docrails project, and tweeting too much. As one of the first members of the RailsBridge initiative (http://railsbridge.com), he tries to expand and enrich the Rails community, one project at a time. He loves to travel and chill out with his two-and-a-half-year-old daughter.

Acknowledgments

Many people helped me with this book project. I would like to thank all of the open source developers who wrote software that I used both for this book and in my work. My wife Carol supported my efforts in many ways, including reviewing early versions to catch typos and offering comments on general readability. My technical editor Peter Szinek made many useful comments and suggestions, as did my editor Michelle Lowman. Project manager Beth Christmas kept me on schedule and ensured that everything ran smoothly. Copy editor Nina Goldschlager Perry helped me improve the general readability of my text. Production editor Ellie Fountain made the final manuscript look good. I would also like to thank Apress staff who helped, even though I did not directly interact with them: indexer Kevin Broccoli, proofreader Liz Welch, compositor Dina Quan, and artist Kinetic Publishing.

Introduction

This book covers Web 3.0 technologies from a software developer's point of view. While non-techies can use web services and portals that other people create, developers have the ability to be creators and consumers at the same time—by integrating their work with other people's efforts.

The Meaning of Web 3.0

Currently, there is no firm consensus on what "Web 3.0" means, so I feel free to define Web 3.0 for the context of this book and to cover Ruby technologies that I believe will help you develop Web 3.0 applications. I believe that Web 3.0 applications will be small, that they can be constructed from existing web applications, and that they can be used to build new web applications. Most Web 3.0 technologies will be important for both clients and services. Web 3.0 software systems will need to find and "understand" information, merge information from different sources, and offer flexibility in publishing information for both human readers and other software systems. Web 3.0 applications will also take advantage of new "cloud" computing architectures and rich-client platforms.

Web 3.0 also means you can create more powerful applications for less money by using open source software, relying on public Linked Data sources, and taking advantage of third-party "cloud" hosting services like Amazon EC2 and Google App Engine.

Reasons for Using Ruby

This book reflects a major trend in software development: optimizing the process by saving programmer time rather than computing resources. Ruby is a concise and effective programming language that I find ideal for many development tasks. Ruby code will probably never run as fast as natively compiled Common Lisp or server-side Java—both of which I also use for development. Ruby hits a sweet spot for me because much of the software that I write simply does not require high runtime performance: web scrapers, text-handling utilities, natural language processing (NLP) applications, system-administration utilities, and low- or medium-volume web sites and web portals.

There are other fine scripting languages. Python in particular is a widely used and effective scripting language that, like Ruby, also finds use in medium- and large-scale systems. The choice of using Ruby for this book is a personal choice. I actually started using Python before Ruby (and used ABC, a precursor to Python, back in ancient history). But once I started using Ruby, I felt that I had happily concluded my personal search for a lightweight scripting language to augment and largely replace the use of Common Lisp and Java in my day-to-day development.

Motivation for Developing Web 3.0 Applications

The world of information will continue to catch up in importance with the physical world. While food, shelter, family, and friends are the core of our existence, we're seeing tighter coupling between the world of information and the physical aspects of our lives. As developers of Web 3.0 technologies and beyond, we have the opportunity to help society in general by increasing our abilities to get the information we need, make optimal decisions, and share with the world both raw information and information-aggregation resources of our own creation.

I consider Web 3.0 technologies to be an evolutionary advance from the original Web and Web 2.0. The original Web is characterized by linked pages and other resources, whereas Web 2.0 is commonly defined by supporting social networks and web-based systems that in general utilize contributions from active users (I would also add the slow integration of Semantic Web technologies to this definition). Only time will tell how Web 3.0 technologies evolve, but my hope is that there will be a balance of support for both human users and software agents—for both consuming and generating web-based information resources.

Evolution of the Web

The evolution of the Web has been greatly facilitated by the adoption of standards such as TCP/IP, HTML, and HTTP. This success has motivated a rigorous process of standardization of Semantic Web technologies, which you will see in Part 2 of this book. Examples from Part 3 take advantage of information resources on the Web that use standards for Linked Data. You will also see the advantages of using standard web-service protocols in Part 4, when we look at techniques for publishing information for both human readers and software agents.

The first version of the Web consisted of hand-edited HTML pages that linked to other pages, which were often written by people with the same interests. The next evolutionary step was database-backed web sites: data in relational databases was used to render pages based on some interaction with human readers. The next evolutionary step took advantage of user-contributed data to create content for other users. The evolution of the Web 3.0 platform will support more automation of using content from multiple sources and generating new and aggregated content.

I wrote this book specifically for software developers and not for general users of the Web, so I am not going to spend too much time on my personal vision for Web 3.0 and beyond. Instead, I will concentrate on practical technologies and techniques that you can use for designing and constructing new, useful, and innovative systems that process information from different sources, integrate different sources of information, and publish information for both human users and software agents.

Book Contents

The first part of this book covers practical techniques for dealing with and taking advantage of rich-document formats. I also present some of the techniques that I use in my work for determining the "sentiment" of text and automatically extracting structured information from text. Part 2 covers aspects of the Semantic Web that are relevant to the theme of this book: discovering, integrating, and publishing information.

Part 3 covers techniques for gathering and processing information from a variety of sources on the Web. Because most information resources do not yet use Semantic Web technologies, I discuss techniques for automatically gathering information from sources that might use custom or ad-hoc formats.

Part 4 deals with large-scale data processing and information publishing. For my own work, I use both the Rails and Merb frameworks, and I will show you how to use tools like Rails and Hadoop to handle these tasks.

Ruby Development

I am assuming that you have at least some experience with Ruby development and that you have a standard set of tools installed: Ruby, irb, gem, and Rails. Currently, Ruby version 1.8.6 is most frequently used. That said, I find myself frequently using JRuby either because I want to use existing Java libraries in my projects or because I want to deploy a Rails web application using a Java container like Tomcat, JBoss, or GlassFish. To make things more confusing, I'll point out that Ruby versions 1.9.x are now used in some production systems because of better performance and Unicode support. I will state clearly if individual examples are dependent on any specific version of Ruby; many examples will run using any Ruby version.

Ruby provides a standard format for writing and distributing libraries: gems. I strongly encourage you to develop the good habit of packaging your code in gem libraries. Because a strong advantage of the Ruby language is brevity of code, I encourage you to use a "bottom-up" style of development: package and test libraries as gems, and build up domain-specific languages (DSLs) that match the vocabulary of your application domain. The goal is to have very short Ruby applications with complexity hidden in DSL implementations and in tested and trusted gem libraries.

I will use most of the common Ruby programming idioms and assume that you are already familiar with object modeling, using classes and modules, duck typing, and so on.

Book Software

The software that I have written for this book is all released under one or more open source licenses. My preference is the Lesser General Public License (LGPL), which allows you to use my code in commercial applications without releasing your own code. But note that if you improve LGPL code, you are required to share your improvements. In some of this book's examples, I use other people's open source projects, in which cases I will license my example code with the same licenses used by the authors of those libraries.

The software for this book is available in the Source Code/Download area of the Apress web site at http://www.apress.com. I will also maintain a web page on my own web site with pointers to the Apress site and other resources (see http://markwatson.com/books/web3_book/).

To make it easier for you to experiment with the web-service and web-portal examples in this book, I have made an Amazon EC2 machine image available to you. You'll learn more about this when I discuss cloud services. Appendix A provides instructions for using my Amazon Machine Image (AMI) with the book examples.

Development Tools

I use a Mac and Linux for most of my development, and I usually deploy to Linux servers. I use Windows when required by customers. On the Mac I use TextMate for writing small bits of Ruby code, but I prefer IDEs such as RubyMine, NetBeans, and IntelliJ IDEA, all of which offer good support for Ruby development. You should use your favorite tools—there are no examples in this book that depend on specific development tools. It is worth noting that Microsoft is making Ruby a supported language on the .NET Framework.

PART 1

■ ■ ■

Text Processing

Part 1 of this book gives you the necessary tools to process text in Web 3.0 applications. In Chapter 1, you'll learn how to parse text from common document formats and convert complex file types to simpler types for easier processing. In Chapter 2, you'll see how to clean up text, segment it into sentences, and perform spelling correction. Chapter 3 covers natural language processing (NLP) techniques that you'll find useful for Web 3.0 applications.

CHAPTER 1

■ ■ ■

Parsing Common Document Types

Rich-text file formats are a mixed blessing for Web 3.0 applications that require general processing of text and at least some degree of semantic understanding. On the positive side, rich text lets you use styling information such as headings, tables, and metadata to identify important or specific parts of documents. On the negative side, dealing with rich text is more complex than working with plain text. You'll get more in-depth coverage of style markup in Chapter 10, but I'll cover some basics here.

In this chapter, I'll introduce you to the `TextResource` base class, which lets you identify and parse a text resource's tagged information such as its title, headings, and metadata. Then I'll derive several subclasses from it to help you parse text from common document formats such as plain-text documents, binary documents, HTML documents, RSS and Atom feeds, and more. You can use the code as-is or modify it to suit your own needs. Finally, I'll show you a couple command-line utilities you can use to convert PDF and Word files to formats that are easier to work with.

Representing Styled Text

You need a common API for dealing with text and metadata from different sources such as HTML, Microsoft Office, and PDF files. The remaining sections in this chapter contain implementations of these APIs using class inheritance with some "duck typing" to allow the addition of plug-ins, which I'll cover in Chapters 2 and 3. If certain document formats do not provide sufficient information to determine document structure—if a phrase is inside a text heading, for example—then the API implementations for these document types simply return no information.

You want to identify the following information for each input document:

- Title

- Headings

- URI

You use the Ruby class `TextResource` to extract information from any text resource that possibly has its title, headings, and metadata tagged. Here is the complete source listing for the `TextResource` class:

```ruby
class TextResource
  attr_accessor :source_uri
  attr_accessor :plain_text
  attr_accessor :title
  attr_accessor :headings_1
  attr_accessor :headings_2
  attr_accessor :headings_3
  attr_accessor :sentence_boundaries
  attr_accessor :categories
  attr_accessor :place_names
  attr_accessor :human_names
  attr_accessor :summary
  attr_accessor :sentiment_rating # [-1..+1]  positive number
                                  # implies positive sentiment

  def initialize source_uri=''
    @source_uri = source_uri
    @title = ''
    @headings_1 = []
    @headings_2 = []
    @headings_3 = []
  end
  def cleanup_plain_text text # just a placeholder until chapter 2
    text
  end
  def process_text_semantics! text # a placeholder until chapter 3
  end
end
```

The most important things for you to notice are the attributes and the two placeholder methods. I'll introduce the attributes in this chapter and delve into them further in Chapters 2 and 3, and I'll implement the cleanup_plain_text and process_text_semantics methods in Chapters 2 and 3, respectively.

■Note The source code for this book contains a single gem library called text-resource that contains all the code for the TextResource class and other examples developed in Chapters 1 through 3. You can find the code samples for this chapter in the Source Code/Download area of the Apress web site (http://www.apress.com).

You will never directly create an instance of the TextResource class. Instead, you will use subclasses developed in the remainder of this chapter for specific document formats (see Figure 1-1). In Chapters 2 and 3, you will "plug in" functionality to the base class TextResource. This functionality will then be available to the subclasses as well.

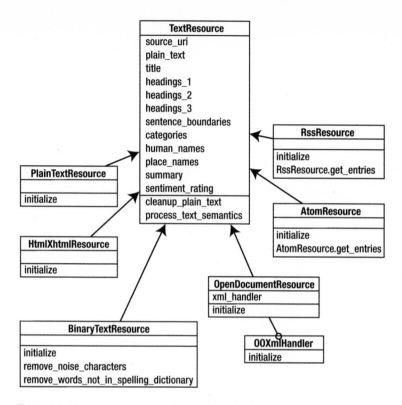

Figure 1-1. *TextResource base class and derived classes*

The RssResource and AtomResource classes (see Figure 1-1) have static class factories for creating an array of text-resource objects from RSS and Atom blog feeds. (You'll learn more about RssResource and AtomResource in the corresponding subsections under the section "Implementing Derived Classes for Different Document Types.")

As a practical software developer, I consider it to be a mistake to reinvent the wheel when good open source libraries are available for use. If existing libraries do not do everything that you need, then consider extending an existing library and giving your changes back to the community. I use the following third-party gem libraries in this chapter to handle ZIP files and to parse RSS and Atom data:

- `gem install rubyzip`

- `gem install simple-rss`

- `gem install atom`

- `gem install nokogiri`

These libraries all work with either Ruby 1.8.6 or Ruby 1.9.1.

Implementing Derived Classes for Different Document Types

In this section, I'll show you the implementations of classes that I'll derive from Ruby's TextResource class, each of which is shown in Figure 1-1. You can use these derived classes to parse data from the corresponding document types.

Plain Text

The base class TextResource is abstract in the sense that it provides behavior and class attribute definitions but does not handle any file types. In this section, I implement the simplest derived class that you will see in this chapter: PlainTextResource. The implementation of this class is simple because it only needs to read raw text from an input URI (which can be a local file) and use methods of the base class:

```
class PlainTextResource < TextResource
  def initialize source_uri=''
    super(source_uri)
    file = open(source_uri)
    @plain_text = cleanup_plain_text(file.read)
    process_text_semantics(@plain_text)
  end
end
```

Except for reading text from a URI (web or local file), all other class behavior is implemented from the TextResource superclass. You can use this PlainTextResource class for any information sources in plain text or for structured data that is externally converted to plain text.

Binary Document Formats

The class you use to parse binary documents differs from the class you use to parse plain-text documents because you need to remove unwanted characters and words. The strategy is to read a binary file as if it were text and then discard nonprinting ("noise") characters and anything that is not in a spelling dictionary. (You'll learn more about noise characters in Chapter 2.)

Here's the code for the BinaryTextResource class, which is also derived from the base class TextResource:

```
class BinaryTextResource < TextResource
  def initialize source_uri=''
    puts "++ entered BinaryPlainTextResource constructor"
    super(source_uri)
    file = open(source_uri)
    text = file.read
    text = remove_noise_characters(text)
    text = remove_words_not_in_spelling_dictionary(text)
    @plain_text = cleanup_plain_text(text)
```

```
    process_text_semantics(@plain_text)
  end
  def remove_noise_characters text
    text # stub: will be implemented in chapter 2
  end
  def remove_words_not_in_spelling_dictionary text
    text # stub: will be implemented in chapter 2
  end
end
```

I'll implement the two stub methods (remove_noise_characters and remove_words_not_
in_spelling_dictionary) in Chapter 2 when I discuss strategies for cleaning up data sources.
(You'll also find the complete implementation in the code samples for this book on the Apress
web site.)

HTML and XHTML

There are several gem libraries for parsing HTML. I use the Nokogiri library in this chapter
because it also parses XML, which means it supports Extensible Hypertext Markup Language
(XHTML). So the example code in this section works for both HTML and XHTML. I will discuss
only the processing of "clean" HTML and XHTML here; Part 3 of the book covers how to process information from web sites that contain advertisements, blocks of links to other web sites
that are not useful for your application, and so on. For those cases, you need to use custom,
site-specific web-scraping techniques.

Before showing you the derived class for parsing HTML and XHTML, I'll give you a quick
introduction to Nokogiri in which I use Nokogiri's APIs to fetch the HTML from my web site
(I'll remove some output for brevity). Here's a snippet from an interactive irb session:

```
irb(main):001:0>   require 'nokogiri'
=> true
irb(main):002:0>   require 'open-uri'
=> true
irb(main):003:0> doc = Nokogiri::HTML(open('http://markwatson.com'))
=> ...
>> doc.class
=> Nokogiri::HTML::Document
irb(main):004:0> (doc.public_methods - Object.public_methods).sort
=> ["/", "<<", "[]", "[]=", "add_child", "add_next_sibling",
"add_previous_sibling", "after", "at", "attributes", "before", "blank?",
"cdata?", "child", "children", "collect_namespaces", "comment?",
"content", "content=", "css", "css_path", "decorate", "decorate!", "decorators",
"document", "document=", "encode_special_chars", "get_attribute",
"has_attribute?", "html?", "inner_html", "inner_text", "internal_subset",
"key?", "name=", "namespaces", "next", "next_sibling", "node_cache",
"parent=", "path", "pointer_id", "previous_sibling", "remove", "remove_attribute",
"replace", "root", "root=", "search", "serialize", "set_attribute", "slop!",
"text", "to_html", "to_xml", "traverse", "unlink", "xml?", "xpath"]
```

I suggest that you try the preceding example yourself and experiment with the methods for the Nokogiri::HTML::Document class listed at the end of the snippet. (I'll show you portions of irb sessions throughout this book.)

In order to extract all of the plain text, you can use the inner_text method:

```
irb(main):005:0> doc.inner_text
=> "Mark Watson, Ruby and Java Consultant and Author\n … "
```

The plain text contains new-line characters and generally a lot of extra space characters that you don't want. In the next chapter, you'll learn techniques for cleaning up this text; for now, the TextResource base class contains a placeholder method called cleanup_plain_text for cleaning text. Nokogiri supports XML Path Language (XPath) processing, DOM-style processing, and Cascading Style Sheets (CSS) processing. I'll start with the DOM (Document Object Model) APIs. I am assuming that you are also using irb and following along, so I am showing only the output for the first child element and the inner text of the first child element:

```
irb(main):006:0> doc.root.children.each {|node| pp node; pp node.inner_text }
#<Nokogiri::XML::Element:0x32f618
 @document=
  #<Nokogiri::HTML::Document:0x104e4c0
   @decorators=nil,
   @node_cache=
    {23674064=>
      #<Nokogiri::XML::Element:0x32f2f8
       @document=#<Nokogiri::HTML::Document:0x104e4c0 ...>>,
     23673296=>#<Nokogiri::XML::Element:0x32f618 ...>,
     23672576=>
      #<Nokogiri::XML::Element:0x32fba4
       @document=#<Nokogiri::HTML::Document:0x104e4c0 ...>>}>>
"Mark Watson, Ruby and Java Consultant and Author\n"
```

As you can see, dealing with HTML using DOM is tedious. DOM is appropriate for dealing with XML data that has a published schema, but the free-style nature of HTML (especially "handwritten" HTML) makes DOM processing difficult.

Fortunately, the XPath APIs are just what you need to selectively extract headings from an HTML document. You use XPath to find patterns in nested elements; for example, you'll use the pattern '//h3' to match all HTML third-level heading elements. Combine XPath with the inner_text method to extract headings:

```
 irb(main):007:0> doc.xpath('//h1')
=> <h1 class="block" align="center">
     Mark Watson: Ruby and Java Consultant and Author
   </h1>
irb(main):008:0> doc.xpath('//h1').inner_text.strip
=> "Mark Watson: Ruby and Java Consultant and Author"
```

By substituting '//h2', '//h3', and '//h4' for the XPath expression, you can collect the page headers. As another example, here's how you would collect all of the headings of level h3:

```
irb(main):009:0> doc.xpath('//h3').collect {|h| h.inner_text.strip}
=> ["I specialize in Java, Ruby, and Artificial Intelligence (AI) technologies",
"Enjoy my Open Content Free Web Books and Open Source Software", "Recent News"]
```

Now you're ready to use the HtmlXhtmlResource class, which is derived from TextResource and included in the text-resource gem library. This is the code for processing HTML and XHTML resources:

```
doc = Nokogiri::HTML(open(source_uri))
@plain_text = cleanup_plain_text(doc.inner_text)
@headings_1 = doc.xpath('//h1').collect {|h| h.inner_text.strip}
@headings_2 = doc.xpath('//h2').collect {|h| h.inner_text.strip}
@headings_3 = doc.xpath('//h3').collect {|h| h.inner_text.strip}
```

The TextResource class's cleanup_plain_text utility method is currently a placeholder; I'll implement it in Chapter 2. Running the preceding code yields these extracted headers from my web site:

```
@headings_1=["Mark Watson: Ruby and Java Consultant and Author"],
@headings_2=["Blogs", "Fun stuff"],
@headings_3=
  ["I specialize in Java, Ruby, and Artificial Intelligence (AI) technologies",
   "Enjoy my Open Content Free Web Books and Open Source Software",
   "Recent News"],
```

Here is the complete class implementation:

```
class HtmlXhtmlResource < TextResource
  def initialize source_uri=''
    super(source_uri)
    # parse HTML:
    doc = Nokogiri::HTML(open(source_uri))
    @plain_text = cleanup_plain_text(doc.inner_text)
    @headings_1 = doc.xpath('//h1').collect {|h| h.inner_text.strip}
    @headings_2 = doc.xpath('//h2').collect {|h| h.inner_text.strip}
    @headings_3 = doc.xpath('//h3').collect {|h| h.inner_text.strip}
    process_text_semantics(@plain_text)
  end
end
```

This code extracts headings based on the heading level of HTML tags.

■**Note** For JRuby developers, I provide example code in the next section for using the pure Ruby REXML library to grab all text (attributes and element text). For processing HTML, you can use a pure Ruby library such as ymHTML (included in the source code for this book on the Apress web site).

OpenDocument

Now I'll discuss the OpenDocumentResource class, which lets you parse text from documents in OpenOffice.org's OpenDocument format. You won't find many web resources in this document format, but it's an international standard that's used by at least five word processors. I include support for OpenDocument in this chapter because this format is ideal for maintaining document repositories. OpenOffice.org offers batch-conversion utilities for converting various Microsoft Office formats and HTML to the OpenDocument format. You can select directories of files for conversion using the application's menus.

The OpenDocument format is an easy-to-read, easy-to-parse XML format that is stored in a ZIP file. First use the standard Ruby ZIP library to extract the ZIP entry named content.xml. Then use the REXML XML parser by providing a Simple API for XML (SAX) XML event handler as a nested class inside the implementation of the OpenDocumentResource class:

```
class OpenDocumentResource < TextResource
  class OOXmlHandler
    include StreamListener
    attr_reader :plain_text
    attr_reader :headers
```

REXML calls the tag_start method for each new starting XML tag:

```
    def tag_start name, attrs
      @last_name = name
    end
```

You need to save the element name so you know what the enclosing element type is when the text method is called. REXML calls the text method whenever text is found in the input stream. The XML for the document content has many elements starting with text:. You'll collect all the inner text from any element whose name contains text:h and save it in an array of header titles. You'll also collect the inner text from any element whose name contains text and save it in the plain-text buffer:

```
    def text s
      s.strip!
      if @last_name.index('text:h')
        @headers << s if s.length > 0
      end
      if @last_name.index('text')
        if s.length > 0
          @plain_text << s
          @plain_text << "\n"
        end
```

```
      end
    end
  end # ends inner class StreamListener
```

The OpenDocumentResource class constructor uses the internal SAX callback class to parse the XML input stream read from the ZIP file entry content.xml:

```
  def initialize source_uri=''
    Zip::ZipFile.open(source_uri) {
      |zipFile|
      xml_h = OOXmlHandler.new
      Document.parse_stream((zipFile.read('content.xml')), xml_h)
      @plain_text = cleanup_plain_text(xml_h.plain_text)
      @headers_1 = xml_h.headers
    }
    process_text_semantics(@plain_text)
  end
end
```

The OpenDocument standard, which is implemented by many word-processing systems, is ideal for creating and maintaining document repositories. Here, you only collected the headers and plain text from OpenDocument files, but the format is richer than the simple Ruby code in the OpenDocumentResource class indicates. If you are interested, I recommend that you try unzipping any OpenDocument file and examine both the metadata and contents-file entries.

I am using OpenOffice.org to write this book. You might find this amusing: I used the OpenDocument file for this chapter as my test data for writing the OpenDocumentResource class.

RSS

Another useful source of information on the Web is web blogs that use either RSS or Atom XML-syndication formats. I originally considered not supporting web blogs as a subclass of TextResource because a single blog URI refers to many blog entries, but I decided to implement RSS and Atom classes with factories for returning an array of blog entries for a single blog URI. These derived classes are called RssResource and AtomResource (see Figure 1-1). This decision makes sense: a static class-factory method returns a collection of TextResource instances, each with the semantic processing performed by the code that you will see in Chapter 3.

The implementation of RSS-feed reading is simple using Lucas Carlson's simple-rss gem library. The simple-rss library handles both RSS 1.0 and RSS 2.0. The RssResource constructor calls the TextResource constructor to initialize instance data to default empty strings and empty lists. The static class method get_entries is a factory that creates an array of RssResource objects from a blog URL:

```
class RssResource < TextResource
  def initialize
    super('')
  end
  def RssResource.get_entries source_uri = ''
    entries = []
```

You never directly create instances of class `RssResource`. You call the static method `RssResource.get_entries` with a web log's source URI, and this static method acts as a factory for returning an array of `RssResource` instances—one instance for each blog entry. The `SimpleRSS.parse` static method returns an instance of class `SimpleRSS`, which is defined in the Ruby simple-rss gem library:

```
rss = SimpleRSS.parse(open(source_uri))
```

The `items` method returns an array of hash tables, each of which contains data for a blog entry:

```
items = rss.items
items.each {|item|
  content = item[:content_encode] || item[:description] || ''
  entry = RssResource.new
  entry.plain_text = entry.cleanup_plain_text(content)
  entry.process_text_semantics(entry.plain_text)
  entry.source_uri = item[:link] || ''
  entry.title = item[:title] || ''
  entries << entry
}
  entries
end
end
```

The `item` method (represented by the block variable `item` in the preceding code snippet) are Ruby hash tables, which have the following keys:

```
irb(main):009:0> entry.keys
=> [:pubDate, :content_encoded, :dc_creator, :description, :link, :title, :guid,
:content]
```

When you use the `RssResource` class's static `get_entries` factory method on the Boing-Boing.net blog, the first returned `RssResource` object looks like this (some output has been removed for brevity):

```
[#<RssResource:0x2b2a49c
  @human_names=["Danny Choo"],
  @place_names=["Tokyo", "Japan"],
  @plain_text=
   "Danny Choo is a guest blogger on Boing Boing. Danny resides in Tokyo,
and blogs about life in Japan and Japanese subculture...",
  @sentence_boundaries=
   [[0, 43],
    [45, 163],    ...
```

This sample output also shows some of the semantic processing that I will add to the TextResource base class in Chapter 3; notice that human and place names are identified.

Atom

Like RSS, Atom is an XML-syndication format. While there is some disagreement over which format is better for our purposes, both formats are easy to work with. You can choose from several Ruby Atom parsers, one of which is the gem atom library. I've been using it for a few years, so I'll use it to implement the class AtomResource. This library was written and is maintained by Brian McCallister and Martin Traverso.

The AtomResource class constructor's initialize method simply calls the superclass constructor with an empty string because the static factory method will insert text into new instances as they are created. The static class method get_entries is a factory method that returns an array of AtomResource instances:

```ruby
class AtomResource < TextResource
  def initialize
    super('')
  end
  def AtomResource.get_entries source_uri=''
    ret = []
    str = Net::HTTP::get(URI::parse(source_uri))
    atom = Atom::Feed.new(str)
    entries = atom.entries
    entries.each {|entry|
```

You use each entry (of type Atom::Entry) to create an instance of AtomResource:

```ruby
      temp = AtomResource.new
      content = entry.content.value || ''
      temp.plain_text = temp.cleanup_plain_text(content)
      temp.process_text_semantics(temp.plain_text)
      temp.source_uri = entry.links[0].href || ''
      temp.title = entry.title || ''
      ret << temp
    }
    ret
  end
end
```

The class Atom::Entry has the following public methods:

```
irb(main):018:0> (item.public_methods - Object.public_methods).sort
=> ["authors", "categories", "content", "contributors", "extended_elements",
"links", "published", "rights", "source", "summary", "title", "updated"]
```

In the AtomResource class, I am using only the accessor methods content, links, and title. Here is an instance of AtomResource created from one of my blog entries (with some output removed for brevity):

```
[#<AtomResource:0x234abc8
  @plain_text=
```

```
"How often do you read something that you totally agree with?
This article by ....",
  @sentence_boundaries=
  [[0, 59], [61, 122], [124, 299], [301, 466], [468, 599], [601, 747]],
  @source_uri=
  "http://www.blogger.com/feeds/7100397/1192914280932390329/comments/default",
  @summary=
  "His take that the Apache 2 (gift), GPL 3 (force people to share), and the
LGPL 3 (an \"in between\" license) that cover the spectrum...",
  @title="Bruce Perens on the GNU Affero General Public License">, ... ]
```

This output shows an automatically generated summary. Summarization is one of the semantic-processing steps that I'll add to the TextResource base class in Chapter 3.

Handling Other File and Document Formats

For my own work, I often use command-line utilities to convert files to formats that are easier to work with—usually plain text, HTML, or OpenDocument. You can also take advantage of several free and commercial utilities to convert between most types of file formats. (In Chapter 9, you'll see that these file conversions prove useful when you want to cache information sources on your local network to implement local search.) I will not derive subclasses from TextResource for PDF and Microsoft Word files because I usually convert these to plain text and use the PlainTextResource class.

Handling PDF Files

I use the pdftotxt command-line utility program to convert PDF files to plain text. For example, you'd type this on the command line if you want to create a text file book.txt from the input file book.pdf:

```
pdftotext -nopgbrk  book.pdf
```

You can find pdftotext at these web sites:

- Compiled Mac binary: http://www.bluem.net/downloads/pdftotext_en/
- Source code and compiled Windows and Linux binaries: http://www.foolabs.com/xpdf/download.html

I find pdftotext to be useful when I do not need to extract the header information from PDF files. If you do need the header information, convert PDF to HTML using pdftohtml and subsequently use HtmlXhtmlResource instead of PlainTextResource to perform your processing. When I need to access the structure of PDF files, I use a Java open source tool such as iText.

When you use pdftohtml, the generated HTML requires some additional processing to extract headers because the generated HTML does not contain <h1>..</h1> style markup. Instead, headers are preceded by an anchor and terminated with a break tag:

```
<A name=1></a>Chapter 1<br>
```

Note that the HTML markup is not well-formed, but you can extract headers using code like this:

```
File.readlines('books.html').map {|line|
  if line.index('<A name=')
    header = line[line.index('</a>')+4...line.index('<br>')]
    puts header if header.length > 2
  end
}
```

This code snippet assumes that the `<a>` and `</a>` tokens appear on the same input line. This is generally not a good assumption for HTML, but the `pdftohtml` utility does place them on one line when it generates HTML text from PDF files.

Handling Microsoft Word Files

I use Adri van Os's `antiword` command-line utility to extract plain text from Microsoft Word documents. You can download the source code or compiled binaries for most operating systems at `http://www.winfield.demon.nl`.

You can use the back-quote character in Ruby to execute an external program and capture any output to a string:

```
puts `antiword test.doc`
```

This code snippet is useful for extracting text from any Microsoft Word document on your file system.

Other Resources

You might find these two additional resources useful in your applications:

- GNU Metadata Extractor Library
- My FastTag Ruby Part-of-speech Tagger

I decided not to include them in the `text-resource` gem library, so I'll provide a brief description of each.

GNU Metadata Extractor Library

The GNU Metadata Extractor Library is released under the GNU general public license (GPL). To install Ruby support for this library, download and install the extractor and then the Ruby bindings from these sites:

- `http://ftp.gnu.org/pub/gnu/libextractor/libextractor-0.5.21.tar.gz`
- `http://extractor.rubyforge.org/`

With this library, you can extract the metadata from most types of files that you find on the web: HTML, PDF, PostScript (PS), OLE2 (DOC, XLS, PPT), OpenOffice (SXW), StarOffice (SDW), Free Lossless Audio Codec (FLAC), MP3, OGG, WAV, EXIV2, JPEG, GIF, PNG, TIFF,

MPEG, QuickTime (QT), Advanced Systems Format (ASF), and so on. Here's a sample `irb` session:

```
irb(main):001:0> require 'extractor'
=> true
irb(main):002:0> Extractor.extract('/Users/markw/Documents/Reading and
Papers/purchased/Programming Erlang.pdf')
=> {"format"=>"PDF 1.4", "creator"=>"LaTeX with hyperref package", "title"=>
"Programming Erlang", "creation date"=>"20070717080117",
"author"=>"The Pragmatic Bookshelf \\(56843\\)", "subject"=>"",
"mimetype"=>"application/pdf", "modification date"=>"D:20070717140124",
"keywords"=>""}
```

FastTag Ruby Part-of-speech Tagger

You can download my open source Part-of-speech Tagger from my web site (`http://markwatson.com/opensource/`) if you need to identify the part of speech for each word in text. Here is a short example:

```
 irb(main):002:0> tt = Tagger.new
=> #<Tagger:0x1033044 @lexicon={"marinating"=>["VBG"], "generalizations"=>["NNS"],
"apocryphal"=>["JJ"], "bid"=>["NN", "VBD", "VBN", "VBP", "VB"], ...
irb(main):003:0> tt.getTags("The bank gave Sam a loan last week.
He can bank an airplane really well.")
=> ["DT", "NN", "VBD", "NNP", "DT", "NN", "JJ", "NN", nil, "PRP", "MD", "NN",
"DT", "NN", "RB", "RB"]
```

I am not using the Part-of-speech Tagger in this book, but if it looks interesting or useful to you, refer to the project's directions and its definitions of tag types (NN is a noun and DT is a determiner, for example).

Wrapup

In this chapter, you learned to use several subclasses derived from the `TextResource` base class to parse text from documents in different formats. `PlainTextResource`, which you use for plain-text documents, reads raw text from an input URI and performs some basic text processing. The `BinaryTextResource` class helps you remove noise characters from documents in binary format, as well as words that don't appear in a spelling dictionary. You learned how to use the `HtmlXhtmlResource` class along with XPath APIs to selectively extract headings from an HTML document.

The `OpenDocumentResource` class lets you work with documents in OpenOffice.org's Open-Document XML format. You use it to parse the XML input stream and collect any headers and plain text that are found. `RssResource` and `AtomResource` let you return an array of text-resource objects from RSS and Atom blog feeds—an array of blog entries for a single blog URI, in other words. Finally, you saw how to use the command-line utilities `pdftotext` and `antiword` to convert PDF and Word files to simpler formats.

I will further extend my `TextResource` class in Chapters 2 and 3 to provide APIs that indicate the following information for individual words or phrases: locations of sentence boundaries, classification tags for text, whether the word is a proper noun (name of a human being or place), a positive or negative "sentiment," and automatically generated short summaries.

■ ■ ■

Cleaning, Segmenting, and Spell-Checking Text

When extracting text from different sources, you commonly end up with "noise" characters and unwanted whitespace. So you need tools to help you clean up this extracted text. For many applications, you'll also want to segment text by identifying the boundaries of sentences and to spell-check text using a single suggestion or a list of suggestions. In this chapter, you'll learn how to remove HTML tags, extract full text from an XML file, segment text into sentences, perform stemming and spell-checking, and recognize and remove noise characters.

I developed Ruby classes and utility scripts in Chapter 1 for processing a variety of document types for different sources of information. In this chapter, I will augment the base class TextResource with behavior that will also function in the derived classes for different document types. I'll start in the next section with techniques to clean up HTML and XHTML text.

Some of the noise that you remove while cleaning up text could potentially be useful for segmenting text. However, the TextResource class's new code from this chapter calls the segmentation APIs *after* it calls the cleanup APIs. This might be an occasion when you'd want to write custom code or modify my example code—I will comment later on why and how you might want to do this for specific applications or text sources.

Removing HTML Tags

I use three general techniques to remove tags from HTML text: using a regular expression, using the HTML::FullSanitizer class found in the Ruby on Rails action_controller gem, and using the sanitize gem. Because the sanitize gem uses the native hpricot gem library (and therefore cannot be used with JRuby), I'll stick with using the Rails libraries because they work with Ruby 1.8.x, Ruby 1.9.x, and JRuby.

■Note You can review the use of regular expressions at http://www.regular-expressions.info/ruby.html and http://www.rubyist.net/~slagell/ruby/regexp.html.

If you are processing well-formed HTML (with no missing tags) or if you are processing XHTML, then using a regular expression to delete tags works fine:

```
irb(main):001:0> html = "<h1>test heading</h1>Test <b>bold</b> text."
=> "<h1>test eading</h1>Test <b>bold</a> text."
irb(main):002:0> html.gsub(/<\/?[^>]*>/, "")
=> "test headingTest bold text."
```

Notice one small problem here: the words "heading" and "Test" run together. An easy solution is simply to replace tags with a space and then convert two adjacent spaces to a single space:

```
irb(main):003:0> html.gsub(/<\/?[^>]*>/, " ").gsub('  ', ' ').gsub('  ', ' ')
=> " test heading Test bold text."
```

Notice that the first call to gsub now converts matching tags to a space instead of a zero-length string. Cleaning up extra whitespace is important for text from a variety of sources in addition to HTML files. The following example uses the FullSanitizer class included with Rails:

```
require 'rubygems'
require 'action_controller'

def cleanup_plain_text text
  def remove_extra_whitespace text
    text = text. gsub(/\s{2,}|\t|\n/,' ')
    text
  end
  text.gsub!('>', '> ')
  if text.index('<')
    text = HTML::FullSanitizer.new.sanitize(text)
  end
  remove_extra_whitespace(text)
end
```

This implementation also removes tab and new-line characters. Here is an example use of cleanup_plain_text:

```
irb(main):010:0> text="<html><head><title>A Test Title</title></head><body>
This is text in an\n\t HTML body element</body></html>"
=> "<html><head><title>A Test Title</title></head><body>
This is text in an\n\t HTML body element</body></html>"
irb(main):011:0> cleanup_plain_text(text)
=> "A Test Title This is text in an HTML body element"
```

Notice in this example that I nest def definitions because only the cleanup_plain_text function uses remove_extra_whitespace. I could have also a procedure object, but I think that nesting def definitions looks better while hiding the visibility of functions and methods that are used only in one place. I've included this cleanup_plain_text function as a utility method in the final implementation of the `TextResource` class that is included in the source code for this book on the Apress web site.

If you would prefer not to have a dependency on Rails in your projects, you can remove tags from HTML text using the `sanitize` gem, which is the third technique I mentioned. But I find it more convenient to stick with the second technique of using the `HTML::FullSanitizer` class in the Ruby on Rails `action_controller` gem. I always have the Rails gems installed on my servers and the MacBook that I use for development, so using the `HTML::FullSanitizer` class means I don't have to install another gem. I can also maintain the ability to use JRuby, which you can't do when using the `sanitize` gem.

Extracting All Text from Any XML File

In this section, I'll show you two utility scripts that pull all text from any arbitrary XML file. Usually you want to process XML with prior knowledge of its schema, but you'll definitely encounter cases where it makes sense simply to pull all text from an XML document in order to make it searchable—if you want to index data, for example. Search results would indicate specific XML files where the text was found.

I usually use Hpricot or Nokogiri to parse XML because these parsers are mostly written in native C for good performance. Another alternative is LibXml Ruby, the Ruby wrapper for the C library `libxml`. An advantage of using REXML (part of the standard Ruby library) is compatibility with JRuby because REXML is implemented in pure Ruby. For completeness, I am providing examples using both REXML and Nokogiri.

Using REXML

I usually use REXML when I don't need high parsing performance—that is, when I am not processing huge XML files. REXML provides both DOM parsing APIs and SAX (event-based) APIs, as you saw in Chapter 1 when you learned to parse OpenOffice.org's OpenDocument files.

Given an XML file or input stream, our algorithm is simple: start with the root element and recursively visit each XML subelement. For each subelement, collect its text and the text from all its attributes. I've called this function `text_from_xml`:

```
require 'rexml/document'
def text_from_xml filename
  str = ''
  doc = REXML::Document.new(File.new(filename))
  doc.each_recursive {|element|
    str << element.text.strip << ' ' if element.text
    str << element.attributes.values.join(' ').strip << ' '
  }
  str
end
```

Here is a sample use of `text_from_xml`:

```
irb(main):021:0> text_from_xml('test.xml')
=> "Mark cooking  hiking "
```

where the contents of the `test.xml` file are:

```
<person name="Mark">
  <hobby>cooking</hobby>
  <hobby>hiking</hobby>
</person>
```

As you can see from the code in the `text_from_xml` function, the value of the variable `doc` is of type `REXML::Document`. Plus, the method `each_recursive` performs a depth-first search of the XML document's DOM tree and calls the code block with each traversed element. The attributes for each element are stored in an instance of the class `REXML::Attributes`. This provides convenience methods such as `[]`, `[]=`, `add`, and so on for accessing individual attributes in this set (`http://www.ruby-doc.org/core/classes/REXML/Attributes.html`). I used the method values to include attribute values in the result string while discarding the attribute names.

Using Nokogiri

The Nokogiri library that you saw in Chapter 1 handles both HTML and XML. The class `Nokogiri::XML::Document` provides a `text` method that returns all of the text in an XML input stream. This process resembles the REXML implementation in the preceding section, but the extracted text from the Nokogiri implementation contains new-line and tab characters. The text also contains redundant adjacent-space characters that you will want to remove. Here's the `text_from_xml` implementation that uses the Nokogiri library:

```
require 'nokogiri'
def text_from_xml filename
  doc = Nokogiri::XML(open(filename))
  doc.text.gsub("\n", ' ').gsub("\t", ' ').split.join(' ')
  ss = ''
  doc.traverse {|n|
    ss << n.content + ' ' if n.class==Nokogiri::XML::Text
    ss << n.attributes.values.join(' ') + ' '  \
            if n.class==Nokogiri::XML::Element
  }
  ss
end
```

Here's a sample use of the Nokogiri version of `text_from_xml` that incorporates the same `test.xml` file from the preceding section:

```
irb(main):049:0> text_from_xml('test.xml')
=> "cooking  hiking  Mark "
```

You should note a few differences between the Nokogiri and REXML examples of `text_from_xml`. First, because I am performing a traversal in the Nokogiri example, the code processes text elements in reverse order. Consequently, the result string from the Nokogiri example contains the extracted text in a different order from the result string in the REXML example. Second, the value of the variable `doc` is of type `Nokogiri::XML::Document` rather than `REXML::Document`. Third, the Nokogiri example uses the Ruby idiom `a_string.split.join(' ')`

to remove extra adjacent spaces in a string, whereas the REXML example does not remove extra adjacent spaces. An alternative approach would be to use a regular expression such as `text.gsub(/\s{2,}|\t/,' ')`.

■**Caution** The class `Nokogiri::XML::Document` has methods `text` and `content`, but I could not use them here because both methods concatenate all text without adding whitespace characters. This means that words do not get separated by spaces.

Segmenting Text

In this section, you'll learn to identify sentence boundaries reliably. Let's assume that the input text has been cleaned so it contains only words, numbers, abbreviations, and punctuation. This segmenting task involves more than merely identifying periods, exclamation marks, or question marks as sentence terminators. A period, for example, can instead appear within an abbreviation, a floating-point number, or a URL.

Here's an example showing why you can't expect perfect accuracy from an automated system:

```
Mr. Johnson and Ms. Smith are getting married.
The bus just left, Mr.
```

In the first line, you can assume that the periods in "Mr." and "Ms." are *not* sentence terminators. But in the second line, the period in the same "Mr." expression *is* a sentence terminator.

You could use machine learning to determine segmentation patterns from sample text with manually identified sentence breaks. But you'll take a simpler approach here: you'll write down a set of rules for sentence segmentation and implement these rules in Ruby code. Specifically, consider periods to be sentence terminators unless one or more of the following conditions are met:

- A period is part of a number (but not the last character in a number)

- A period is part of one of the following tokens: Mr., Mrs., Ms., Dr., Sr., Maj., St., Lt., or Sen.

Before you begin implementing this sentence-segmentation scheme, you need to take care of an incompatibility between Ruby 1.8.x and Ruby 1.9.x: Ruby 1.8.x has no `each_char` method in the `String` class. The following bit of code adds this method if required:

```
begin
  "1".each_char {|ch| } # fails under Ruby 1.8.x
rescue
  puts "Running Ruby 1.8.x: define each_char"
  class String
    def each_char(&code)
```

```
      self.each_byte {|ch| yield(ch.chr)}
    end
  end
end
```

When I write Ruby libraries, I don't usually use monkey patching (modifying the runtime code). But because I like to use the each_char method, it makes sense here to "update" Ruby 1.8.x. Later in this section, I will also monkey-patch the class String to add the method each_char_with_index_and_current_token. I prefer to separate code from data, so start by defining two arrays of constants that you'll need, one for abbreviations preceding human names and one for floating-point numbers:

```
HUMAN_NAME_PREFIXES = ['Mr.', 'Mrs.', 'Ms.', 'Dr.', 'Sr.', 'Maj.', 'St.',
                                      'Lt.', 'Sen.']
DIGITS = ['0', '1', '2', '3', '4', '5', '6', '7', '8', '9']
```

Given this data, the code is fairly simple. You use the String#each_char method to iterate over each character in the input string. As you iterate over the characters, you maintain a string of the current token to compare against abbreviations such as "Dr." to determine when period characters do not mark the end of a sentence.

The following code is a first-cut implementation of the get_sentence_boundaries method that iterates over the characters in a string while maintaining the current character index in the string and the value of the current token. Tokens are defined here as substrings delineated by space characters:

```
def get_sentence_boundaries text
  boundary_list = []
  start = index = 0
  current_token = ''
  text.each_char {|ch|
    if ch == ' '
      current_token = ''
    elsif ch == '.'
      current_token += ch
```

You use this get_sentence_boundaries method to calculate sentence boundaries and an accessor method called attr_accessor :sentence_boundaries to access the calculated values. Before considering that a period character is a sentence delimiter, you want to confirm that it's not part of a name-prefix abbreviation or a floating-point number:

```
    if !HUMAN_NAME_PREFIXES.member?(current_token) &&
        !DIGITS.member?(current_token[-2..-2])
```

The array boundary_list contains two element arrays, where the first element is the starting character index for a sentence, and the second element is the index of the last punctuation character in a sentence:

```
      boundary_list << [start, index]
      current_token = ''
      start = index + 2
    else
```

```
        current_token += ch
      end
    elsif ['!', '?'].member?(ch)
        boundary_list << [start, index]
        current_token = ''
        start = index + 2
    else
      current_token += ch
    end
    index += 1
  }
  boundary_list
end
```

The following example shows how to use the sentence_boundaries accessor method:

```
require 'pp'
text = "I saw Dr. Jones yesterday. He was at the market paying $2.25 for aspirin.
He got a good deal!"
breaks = sentence_boundaries(text)
pp breaks
breaks.each {|start, stop| puts text[start..stop]}
```

Here's the corresponding output:

```
[[0, 25], [27, 72], [74, 92]]
I saw Dr. Jones yesterday.
He was at the market paying $2.25 for aspirin.
He got a good deal!
```

Because this is a book about both Ruby programming and Web 3.0 development, I'll use some Ruby magic to improve this implementation that uses the method each_char. The problem is that the get_sentence_boundaries code that counts the character index and maintains the current token's value is tedious. Instead, try using the each_char_with_index_ and_current_token method. As I hinted earlier, I'll show you how to extend the String class to add the method each_char_with_index_and_current_token. Start with these lines:

```
class String
  def each_char_with_index_and_current_token()
```

Here I'm using a Ruby trick: the lazy initialization of class attributes. The standard String class does not provide behavior for keeping track of the current token when traversing a string using each_char. I will add to the String class the @current_token and @index attributes, which I will use to track the characters in the current token and the index of the current character as the string is processed. The following two statements only add these two class attributes if they are not already defined:

```
    @current_token ||= ''
    @index ||= 0
```

When you iterate over the characters in a string, keep the class attributes @current_token and @index up-to-date. This way, when you yield to a block of code, the index and current-token values can be passed to the block in addition to the current character:

```
    self.each_char {|ch|
      if ch == ' '
        @current_token = ''
      else
        @current_token += ch
      end
      @index += 1
      yield(ch, @index, @current_token)
    }
  end
end
```

■**Tip** When you use Ruby monkey patching, you unfortunately do not get to make a decision about the visibility of the patch. It would be useful to be able to limit patches to a single method, a single source file, or a single class hierarchy, but this is not possible. In this case, because you used lazy initialization of class attributes, all the program's string objects do *not* automatically have these two new attributes. Only string objects that use the each_char_with_index_and_current_token method called get these two new attributes defined. This is the effect that you want.

With the addition of the string method each_char_with_index_and_current_token, the implementation of the function sentence_boundaries is much simpler to read and understand:

```
def sentence_boundaries text
  boundary_list = []
  start = index = 0
  current_token = ''
  text.each_char_with_index_and_current_token {|ch, index, current_token|
    if ch == '.'
      if !HUMAN_NAME_PREFIXES.member?(current_token) &&
         !DIGITS.member?(current_token[-2..-2])
        boundary_list << [start, index]
        start = index + 1
      end
    elsif ['!', '?'].member?(ch)
        boundary_list << [start, index]
        start = index + 1
    end
  }
  boundary_list
end
```

In addition to the benefit of code readability, splitting out code of a "different theme" into a new method offers advantages such as code reuse and the elimination of coding errors. As a matter of good design technique, splitting out the each_char_with_index_and_current_token code into a new method makes sense because the code that was factored out has nothing specifically to do with the purpose of the sentence_boundaries function that calls it.

Stemming Text

Stemming is the process of simplifying words by removing suffixes, thereby mapping many different but similar words into the same root stem. Matt Mower and Ray Pereda have provided a Ruby gem that you can install in the usual way: gem install stemmer. This gem implements the Porter stemming algorithm that is widely used and implemented in many programming languages. You might want to investigate different strategies for stemming depending on your needs—if the text you are processing, for example, contains special acronyms that you want to preserve.

This gem defines the module Stemmable and includes this module in the String class. A short example will show you what stemming does to a few common words:

```
>> require 'stemmer'
=> true
>> "banking".stem
=> "bank"
>> "tests testing tested".split.collect {|word| word.stem}
=> ["test", "test", "test"]
```

Stemming sometimes simplifies words to shorter character strings that are no longer valid words. If you notice this, don't worry; it is not a bug in the library! I'll explain stemming further in Chapters 3 and 9.

Spell-Checking Text

Peter Norvig published a 19-line Python script that performs spelling correction (http://norvig.com/spell-correct.html), and Brian Adkins translated it into Ruby. You can find both scripts using a web search. Although I have used Peter Norvig's algorithm on a few projects, I usually prefer to use what I consider to be a state-of-the-art spelling system: GNU Aspell. Aspell does not have any Ruby bindings, but I'll give you a script that runs the Aspell system as an external program. My script parses out suggested spelling corrections from the output of the external process.

To use the script, you need to have Aspell installed on your system. (If you don't, it might be simpler to use Peter Norvig's or Brian Adkin's script.) However, Aspell is easy to install on Linux and Mac OS X, and there are also Windows binary installers available. Before I show you my script, I'll run Aspell interactively so you can see the format of its output:

```
markw$ echo "tallkiing" |  /usr/local/bin/aspell -a list
@(#) International Ispell Version 3.1.20 (but really Aspell 0.60.5)
& tallkiing 28 0: talking, tallying, tackling, taking, tallyhoing, stalking,
```

Tallinn, tacking, tailing, telling, tilling, tolling, balking, tanking, tasking, walking, Tolkien, alleging, baulking, caulking, dallying, laking, liking, tailoring, tellering, lacking, larking, tickling

You see that the suggested corrections for "tallkiing" include "talking," "tallying," "tackling," and so on. You'll want the first suggestion in this case.

My script for using Aspell in Ruby includes the get_spelling_correction_list function, which gets a list of suggestions, and the get_spelling_correction function, which gets only the most likely spelling correction:

```ruby
def get_spelling_correction_list word
  aspell_output = `echo "#{word}" | /usr/local/bin/aspell -a list`
  aspell_output = aspell_output.split("\n")[1..-1]
  results = []
  aspell_output.each {|line|
    tokens = line.split(",")
    header = tokens[0].gsub(':','').split(' ')
    tokens[0] = header[4]
    results <<
      [header[1], header[3],
        tokens.collect {|tt| tt.strip}] if header[1]
  }
  begin
    return results[0][2][0..5]
  rescue ; end
  []
end

def get_spelling_correction word
  correction_list = get_spelling_correction_list(word)
  return word if correction_list.length==0
  correction_list[0]
end
```

Here's a sample use of this script that gets corrections for different misspellings of "walking":

```ruby
>> require 'spelling'
=> true
>> require 'pp'
=> false
>> pp get_spelling_correction_list("waalkiing")
["walking", "walling", "weakling", "waking", "waling", "welkin"]
=> nil
>> pp get_spelling_correction("wallkiing")
"walking"
=> nil
```

Users of Web 3.0 web portals will expect automated spelling correction and/or spelling hints. You can add your own application-specific spelling dictionaries to Aspell by following the directions in the Aspell documentation (http://aspell.net).

Recognizing and Removing Noise Characters from Text

In this section, I'll show you how to remove valid text from binary files. If document files are properly processed, you shouldn't get any noise characters in the extraction. (Examples of noise characters include binary formatting codes used in old versions of a word processor that you no longer have access to.) However, it is a good strategy to have tools for pulling readable text from binary files and recovering text from old word-processing files.

Another reason you'd want to extract at least some valid text from arbitrary binary files is if you must support search functionality. For example, if you need to provide a search engine for your local network at work, you might want to let users find arbitrary old documents regardless of their format. I'll show you a trick for extracting valid text from binary files. It often works because word-processing documents usually encode text in ASCII or Unicode Transformation Format (UTF) and intersperse it with control characters containing formatting information.

The source code for this book (downloadable from the Apress web site) contains a sample file called noise.txt that I'll use as an example. I created the contents of this file by grabbing snippets from three different binary files from different word processors:

```
<B6><E7><A1>&<A4><E8>
1 0 0 1 25 778 Tm
-0.0675 Tc
0.4728 Tw
(your physician or any information contained on or in any product ) Tj
ET
Q
q
   <rdf:li rdf:parseType="Resource">
     <stMfs:linkForm>ReferenceStream</stMfs:linkForm>
   </rdf:li>
ASRvtOpXp&#xA;giCZBEY8KEtLy3EHqi4q5NVVd+IH+qDXDkxkHkyO6jdaxJJcpCUYQTkoZ
CDsD3pO3plkMVxJvdQK&#xA;
TXSkUGaCONfTShqNieQ9+vXISs7sJmmC/mNogVWvYAx4OLAOoabd
  Heading 3$???<@&?CJOJQJ?^JaJ<A@???<
    R?h?l
In an effort to ensure timely
bjbjUU  "7|7|???????l??????????
```

As you can see, the file includes some valid text along with a lot of "noise" or "garbage" characters. In this section, let's assume that Ruby 1.8.x has been monkey-patched to include the String::each_char method that you implemented earlier in this chapter. You'll start processing this "noisy" text by removing all characters that are not numbers, letters, or punctuation. Here's a first cut at the remove_noise_characters function, which lets you accomplish this:

```
def remove_noise_characters text
  def valid_character ch
    return true if ch >= 'a' and ch <= 'z'
    return true if ch >= 'A' and ch <= 'Z'
    return true if ch >= '0' and ch <= '9'
    # allow punctuation characters:
    return true if [' ','.',',',';','!'].index(ch)
    return false
  end
  ret = ''
  text.each_char {|char|
    ret <<  (valid_character(char) ? char  : '')
  }
  ret.split.join(' ')
end
```

The nested valid_character function returns true if the input character is one of the characters that you want to save. When you use the preceding code to process the noisy text in the noise.txt file, you get this output:

```
"B6 E7 A1 A4 E8 1 0 0 1 25 778 Tm 0.0675 Tc 0.4728 Tw your physician or any
information contained on or in any product Tj ET Q q rdf li rdf parseType Resource
stMfs linkForm ReferenceStream stMfs linkForm rdf li ASRvtOpXp
xA;giCZBEY8KEtLy3EHqi4q5NVVd IH qDXDkxkHkyO6jdaxJJcpCUYQ
TkoZCDsD3pO3plkMVxJvdQK xA; TXSkUGaCONfTShqNieQ9 vXISs7s
JmmC mNogVWvYAx4OLAOoabd Heading 3
CJOJQJ JaJ A R h l In an effort to ensure timely bjbjUU 7 7 l"
```

As you can see from this result, you still have work to do. The next step is to discard all string tokens that are not in a spelling dictionary. A disadvantage of this approach is that extracted text will not contain "words" that are product numbers, product names, and the like. This is a real shortcoming if the extracted text is indexed for a search engine; a user searching for a product name, for example, probably won't get any search results. One application-specific way to work around this problem is to include application-specific names in a custom word dictionary. For our purposes, a spelling dictionary is a large text file from which you will extract all unique words.

I'll give you two different scripts: a revised version of remove_noise_characters, which removes all garbage and control characters, and remove_words_not_in_spelling_dictionary, which filters out all words that are not in a custom spelling dictionary. Suppose you want to set up an indexing and search system (as you'll be doing in Chapter 9) and that you want to index a wide variety of word-processing documents. In that case, you will certainly want to use the first script to preprocess binary files, and you might also want to use the second script. You'll base this decision on whether you want all possible search words or tokens in an index, or just common words.

The previous code snippet includes a nested function definition valid_character to discard everything but letters, numbers, and a few punctuation characters. Here's a revised version of the first script, remove_noise_characters, with some output abbreviated:

```
>> require 'remove_noise_characters'
Running Ruby 1.8.x: define each_char
=> true
>> str = File.open("noise.txt").read
=> "<B6><E7><A1>&<A4><E8>\n1 0 0 1 25 778 Tm\n-0.0675 ..."
>> s1 = remove_noise_characters(str)
=> "B6 E7 A1 A4 E8 1 0 0 1 25 778 Tm 0.0675 Tc 0.4728 Tw your physician or any
information contained on or in any product Tj ET Q q rdf li rdf parseType Resource
 stMfs linkForm ReferenceStream stMfs linkForm rdf li ASRvtOpXp
xA;giCZBEY8KEtLy3EHqi4q5NVVd IH
qDXDkxkHky06jdaxJJcpCUYQTkoZCDsD3pO3plkMVxJvdQK xA;
TXSkUGaCONfTShqNieQ9 vXISs7sJmmC mNogVWvYAx40LAOoabd Heading
3 CJOJQJ JaJ A R h l In an effort to ensure timely bjbjUU 7 7 l"
```

The second script, remove_words_not_in_spelling_dictionary, requires a large text file containing common words to form a custom dictionary. I use a concatenation of public-domain books and technical information pulled from the Web. The text file you use should be tailored to your application; for example, if you are writing a web portal dealing with agriculture, then you would want to use sources of information that contain common terms used in this domain.

I use the String#scan method to filter out unwanted characters. For example, here I filter out capital letters and punctuation, leaving only lowercase letters:

```
>> "The dog chased the cat.".scan(/[a-z]+/)
=> ["he", "dog", "chased", "the", "cat"]
```

You'll use the String#downcase method to convert a dictionary file to all lowercase letters, which is a good idea for applications that do document clustering and automatic categorization, for example. In addition to checking if a word token was in the original large text file for the custom dictionary, I also want to filter out every single letter token but "a," "i," and a few other common noise tokens:

```
VALID_WORD_HASH = Hash.new
words = File.new('/Users/markw/temp/big.txt').read.downcase.scan(/[a-z]+/)
words.each {|word| VALID_WORD_HASH[word] = true}
 TOKENS_TO_IGNORE = ('a'..'z').collect {|tok| tok if !['a', 'i'].index(tok)} +
                                        ['li', 'h1', 'h2', 'h3', 'h4','br'] - [nil]

def remove_words_not_in_spelling_dictionary text
  def check_valid_word word
    return false if  TOKENS_TO_IGNORE.index(word)
    VALID_WORD_HASH[word] || ['.',';',',','].index(word)
  end
```

```
  ret = ''
  text.gsub('.',' . ').gsub(',', ' , ').gsub(';', ' ; ').split.each {|word|
    ret << word + ' ' if check_valid_word(word)
  }
  ret.gsub('. .', '.').gsub(' ,', ',').gsub('; ;', ';')
end
```

In the first script, the variable s1 contains the value of applying the function remove_ noise_characters to the sample noise.txt file. Now you'll apply the filter that removes words not found in our custom dictionary:

```
>> remove_words_not_in_spelling_dictionary(s1)
=> ". your physician or any information contained on or in any product ;
an effort to ensure timely "
```

Note that removing words not in a spelling dictionary might be a bad idea for some applications. If you are processing text with product IDs, for example, you'd probably want to keep the IDs for searching, document clustering, and so on.

Custom Text Processing

Much of my work in the last ten years has involved text and data mining. If you spend time writing software for text mining, you will soon understand the need for customizing algorithms (and their implementations) for different information sources and different types of text. The code examples that I developed in this chapter are a good introduction and will get you started. There are two basic techniques for developing automated text-processing tools:

- Manually analyze samples from the data sources you need to use and then code hand-crafted solutions based on your observations.

- Collect a large set of training data from your data sources and manually perform desired tagging and/or cleanup tasks. Use machine-learning techniques such as generating rules and decision trees to process text.

The first strategy is much easier, so I recommend you start there. If you need to use the second strategy (which is beyond the scope of this book), I suggest the following tools: Weka, a Java data-mining tool, and Natural Language Toolkit (NLTK), a Python-based tool for natural language processing.

You have already seen the removal of HTML tags from text. You will find that when you process text extracted from some document sets, you might need to remove common noise/ word sequences such as page numbering, footnote references, and so on.

Wrapup

In this chapter, you learned techniques to remove tags from HTML text, pull all text from any arbitrary XML file, and reliably identify sentence boundaries. You also learned how to accomplish stemming, the process of simplifying words by removing suffixes. With regard to spell-checking, I provided a script for using GNU Aspell in Ruby, which includes functions for getting a list of suggestions and getting only the most likely spelling correction. I showed you how to remove invalid text from binary files, process "noisy" text by removing unwanted characters, and discard all string tokens that are not in a spelling dictionary.

The Ruby `text-resource` gem that I've included with source code for this book (downloadable from the Apress web site) integrates the cleanup and sentence-segmentation code snippets and methods that were developed in this chapter. Please feel free to use the code as a starting point for further customization.

Next, you'll learn how to automatically categorize and summarize text, as well as how to determine the "sentiment" of text.

CHAPTER 3

■ ■ ■

Natural Language Processing

Natural language processing (NLP), also known as computational linguistics, is a broad subject that encompasses technologies for automated processing of natural (human) language. Such processing includes parsing natural-language text, generating natural language as a form of program output, and extracting semantics. I have been working in this field since the 1980s using a wide variety of programming techniques, and I believe that quantitative methods provide better results with less effort for most applications. Therefore, in this chapter, I will cover *statistical NLP*. Statistical NLP uses statistical or probabilistic methods for segmenting text, determining each word's likely part of speech, classifying text, and automatically summarizing text. I will show you what I consider to be some of the simplest yet most useful techniques for developing Web 3.0 applications that require some "understanding" of text. (Natural-language generation is also a useful topic, but I won't cover it here.)

■**Note** All the examples in this chapter assume that you are in the directory src/part1 and running an interactive irb session. Whenever I refer to example file names, they are in this directory. You can find the code samples for this chapter in the Source Code/Download area of the Apress web site (http://www.apress.com).

Figure 3-1 shows the main Ruby classes that I'll develop in this chapter. I'll also provide other useful code snippets that aren't shown in Figure 3-1, such as Ruby code that uses the Open Calais web service to find properties and values in text (see the section "Performing Entity Extraction Using Open Calais"). I'll end this chapter by integrating most of the new NLP utilities into the TextResource#process_text_semantics method of the TextResource base class and developing two sample applications.

■**Tip** Most of the examples in this book are installed and ready to run on an Amazon Elastic Compute Cloud (Amazon EC2) Amazon Machine Image (AMI). See Appendix A for directions.

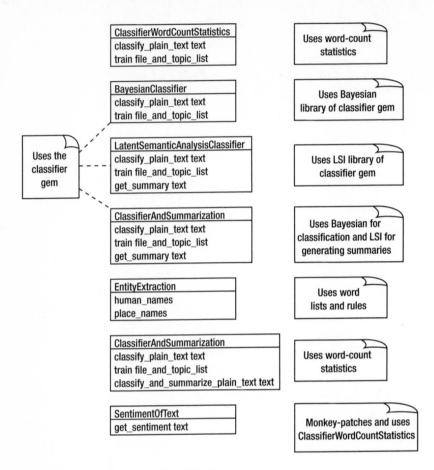

Figure 3-1. *Ruby classes developed in this chapter*

In addition to my own utility classes and the Ruby code that uses the Open Calais web service, I will use the Ruby gem classifier developed by Lucas Carlson and David Fayram. You can install this gem using gem install classifier.

Automating Text Categorization

Automating categorization of text entails automatically assigning tags to text or classifying the content within text. Examples of categories, classifications, or tags might include "politics," "economy," "health," and so on. The categorization-preparation technique I use most often in my work involves compiling word-count data: the number of times each word appears in a document. I then use this data to implement automatic categorization. Here are three steps you can follow to prepare word-count data:

1. Specify the categories (or tags) that your application requires. (I'll use four broad categories in this chapter's examples: Computers, Economy, Health, and Software.)

2. For each category (or tag), collect a set of training documents that "belong" to the category.

3. For each category, calculate a weighted word list that represents the category.

I collect training documents from Wikipedia and other articles I find on the Web. In the following subsections, I will use both my own code and the Ruby gem classifier. In all cases, we are relying on the fact that all the classes to which we are sending messages respond to the `classify_plain_text` method (a process of dynamic typing called "duck typing").

Using Word-Count Statistics for Categorization

The automatic-summarization algorithm that I'll illustrate is dependent on the word-count statistics classifier, which I'll show you how to implement. The first step in building a classification system to assign categories to text is to collect representative text for each category that you want to identify. For the purposes of this chapter, I'll use Wikipedia articles that I selected for the following categories:

- Computers
- Economy
- Health
- Software

For these examples, I'll use the text files located in the `src/part1/wikipedia_text` directory (you can download the source code for this chapter from the Apress web site). I refer to these files as "training data." In addition to learning programming techniques for statistical NLP, your goal is also to extend the base Ruby class `TextResource` that you studied in Chapter 1. In this case, you are training classifiers only for these four sample categories, but for your own applications you will define appropriate categories and collect training text for those categories.

First, you'll extend the base `TextResource` class with an implementation of the class `ClassifierWordCountStatistics`, which reads the training data for each category and builds word-count hash tables:

```
class ClassifierWordCountStatistics
  def initialize
    @category_names = []
    @category_wc_hashes = []
```

In this class implementation, word counts for each category hash table are normalized by the number of words in the training data for a category; this is required so that the training-data sets for each category can contain different amounts of text. The class accomplishes normalization by dividing the word counts by the number of words in the source document. This normalization allows you to properly handle input texts containing different numbers of words.

I am using a short list of noise (or "stop") words to ignore:

```
@noise_words = ['the', 'a', 'at',  ...]  # all entries not shown
  end
```

You can also find long lists of stop words on the Web (at this site, for example: http://www.dcs.gla.ac.uk/idom/ir_resources/linguistic_utils/stop_words).

The classify_plain_text method of ClassifierWordCountStatistics classifies text (that is, assigns a tag or category to it) by finding the category that is associated with words appearing in the input text:

```
def classify_plain_text text
  word_stems = text.downcase.scan(/[a-z]+/)
  scores = []
  @category_names.length.times {|i|
    scores[i] = score(@category_wc_hashes[i], word_stems)
  }
  @category_names[scores.index(scores.max)]
end
```

As you can see from the code, the classify_plain_text method converts text to an array of words that have been converted to lowercase. The temporary variable scores is used to store the calculated score of each trained category.

The following method reads a set of files for each topic (or category) and calculates a normalized word count for how often a word is associated with a category:

```
def train file_and_topic_list
  file_and_topic_list.each {|file, category|
    words = File.new(file).read.downcase.scan(/[a-z]+/)
    hash = Hash.new(0)
    words.each {|word| hash[word] += 1 ➠
        unless @noise_words.index(word) }
    scale = 1.0 / words.length
    hash.keys.each {|key| hash[key] *= scale}
    @category_names << category
    @category_wc_hashes << hash
  }
end
private
def score (hash, word_list)
  score = 0
  word_list.each {|word|
    score += hash[word]
  }
  1000.0 * score / word_list.size
end
end
```

The accuracy of using word-count statistics depends on the coverage of the training data for each category. In this example, I have used arbitrary linear scaling of word counts (that is, I normalize by the number of words in input texts), and the classify_plain_text method returns relative ranking. The example in the next section uses Bayes' theorem to calculate the true probability of whether a document belongs to a given category, dependent on the training data. In practice, you will want to have as much training text as possible for each category you are using in your application.

■Note I will rewrite this class in the section "Automatically Generating Summaries." This rewrite is for efficiency: we need to do much of the same work for classification as summarization.

Here is an example using the `ClassifierWordCountStatistics` class:

```
>> require 'classifier_word_count_statistics'
=> true
>> test = ClassifierWordCountStatistics.new
=> #<ClassifierWordCountStatistics:0x1b4e618   ... >
>> test.train([['wikipedia_text/computers.txt', "Computers"],
?>          ['wikipedia_text/economy.txt', "Economy"],
?>          ['wikipedia_text/health.txt', "Health"],
?>          ['wikipedia_text/software.txt', "Software"]])
=> [["wikipedia_text/computers.txt", "Computers"], ["wikipedia_text/economy.txt",
"Economy"], ["wikipedia_text/health.txt", "Health"],
["wikipedia_text/software.txt", "Software"]]
>> puts test.classify_plain_text(
"Heart attacks and strokes kill too many people every year.")
Health
```

You'll see more examples using this class in the section "Automatically Generating Summaries," after I monkey-patch this class to support generating summaries on input text.

Using a Bayesian Classifier for Categorization

In addition to using word-count statistics, I also sometimes use a Bayesian classifier to automatically categorize text. Using word-count statistics provides a relative measure of similarity, while a Bayesian classifier provides probabilities that a sample of text is assigned to a category.

Bayes' theorem provides a relationship between the probabilities of two events. The following formula:

$$\Pr(A \mid B) = \frac{\Pr(B \mid A) \Pr(A)}{\Pr(B)}$$

states that the probability of A given that B is true is equal to the probability of B given that A is true multiplied by the probability of A divided by the probability of B. This theorem proves useful in situations where you know or you can calculate the probability of B given that A is true.

The classifier gem classifies text by implementing the Naïve Bayes algorithm, which is a simple probabilistic classifier based on Bayes' theorem. In making statements about the existence of attributes for a class, the Naïve Bayes algorithm ignores the existence of other attributes. For our use, this means that you can ignore any effect of other words in text when determining the effect of a word on determining a category. A concrete example: the effect of the word "dog" in determining a category is independent of whether the input text contains the word "cat."

Each category that you want to recognize is represented by a count of the number of occurrences of each word in each training document. For a new sample of text, you want to be able to calculate the probability that this new text appears in each of the trained categories. Using Bayes' theorem, here's how you can calculate the probability that text is in a category:

- Count the number of words in the training text for this category and store the value in the variable count

- Iterate over each word in the new text (store in the variable word)

  - Store the value in a word-count hash (the variable val)

  - Calculate the probability for text being in a category: category += log(val / count)

This algorithm is called "naïve" because words in text are independent of one another. This is not a valid assumption, but the cost of including the effects of word adjacency is high: training takes much longer and requires much more memory, and using a trained model for classification is more expensive computationally. (You can model the effects of word adjacency using Hidden Markov Models, which is beyond the scope of this book.)

The Naïve Bayes algorithm is similar to my word-frequency algorithm (my word-count statistics classifier) if you replace this:

```
score for category += log(val / count)
```

with this:

```
score for category += val / count
```

The source code for this implementation is in the bayes.rb file in the classifier gem.

Note You should be aware of where the source code for installed Ruby gems "lives" in your local file system. I set up a project in my text editor (TextMate on my MacBook) that includes the top-level directory for my installed Ruby gems. (This is in the directory /Users/mark/bin/ruby/lib/ruby/gems/1.8/gems on my system because I like to use private Ruby installations from source code rather than what's bundled with the operating system.) You can see where the installed gems are stored on your computer using the command gem environment. When I am working with a Ruby gem, I often find it useful to open the gem directory in an editor for reference.

The APIs for using the classifier gem are easy to use directly, but here I am wrapping this gem so that the APIs all supply the same public methods for using my classification and summarization libraries. The following snippet shows the wrapper class BayesianClassifier:

```ruby
require 'classifier'

class BayesianClassifier
  def classify_plain_text text
    @bayes.classify(text)
  end
  def train file_and_topic_list
    topic_names = file_and_topic_list.collect {|x| x.last}
    @bayes = Classifier::Bayes.new(*topic_names)
    file_and_topic_list.each {|file, category|
      text = File.new(file).read
      @bayes.train(category, text)
    }
  end
end
```

■**Note** You can use the same instance of `BayesianClassifier` by retraining it with different data. Calling the method `classify_plain_text` before calling the method `train` will cause a runtime error.

Here is a simple example of how you'd use this wrapper class:

```
>> require 'classifier_bayesian_using_classifier_gem'
Notice: for 10x faster LSI support, please install http://rb-gsl.rubyforge.org/
"Health"
=> true
>> test = BayesianClassifier.new
=> #<BayesianClassifier:0x1ba1ee4>
>> test.train([['wikipedia_text/computers.txt', "Computers"],
?>          ['wikipedia_text/economy.txt', "Economy"],
?>          ['wikipedia_text/health.txt', "Health"],
?>          ['wikipedia_text/software.txt', "Software"]])
=> [["wikipedia_text/computers.txt", "Computers"], ["wikipedia_text/economy.txt",
"Economy"], ["wikipedia_text/health.txt", "Health"],
["wikipedia_text/software.txt", "Software"]]
>> puts test.classify_plain_text("
Heart attacks and strokes kill too many people every year.")
"Health"
```

The classifier gem also includes an implementation of Latent Semantic Indexing (LSI) for classification and generating summaries, which I'll demonstrate in the next section.

Using LSI for Categorization

LSI, also commonly known as Latent Semantic Analysis (LSA), takes into account relationships between words found in documents. As a concrete example, suppose you have a training
category called "information for flying airplanes." If you use the Naïve Bayes algorithm that
you used in the last section for classifying text, the existence of the words "bank" ("to bank the
airplane"), "airplane," and "fly" would all be independent of one another as far as determining
if an input text is in the "information for flying airplanes" category.

With LSI, on the other hand, it is likely that the three words "bank," "airplane," and "fly"
would often occur together in documents in the "information for flying airplanes" category. In
a sense, the existence of all three words in one document would reinforce one another's effect
on determining the category "information for flying airplanes."

Another advantage of LSI is calculating the effect of a word in determining a text category
when the word has several different uses. For example, in addition to serving as a verb in the
phrase "bank the airplane," "bank" can serve as a noun in the sense of "depositing money in
the bank" or "fishing on the river bank." In this example, the occurrence of the words "bank"
and fly" in the same document would reinforce the sense of "bank" as a verb meaning to turn
an airplane.

In my tests, the LSI classification code in the classifier gem provides less accurate results
than either my word-count statistics script or the Bayesian classifier in the classifier gem.
However, the LSI library does provide other interesting features such as performing summarization and finding similar document texts.

The LSI library in the classifier gem can run as pure Ruby code or optionally use the GNU
Scientific Library (GSL) to greatly reduce the runtime of some operations. If you make extensive use of the LSI library, I strongly recommend installing the GSL and the GSL gem. This is
easy to do in Linux, a little more difficult to do in Mac OS X, and difficult to do in Windows.
You might want to start using the LSI library in the classifier gem without the GSL installed.

LSI analysis identifies key terms in documents and builds a sparse matrix with dimensions
for terms and documents. If you collect a unique list of all words in all documents being analyzed, then this sparse matrix might have one row for each word in this list and one column
for each document being analyzed. An element in this two-dimensional matrix is indexed by
a specific term and a specific document, and indicates if the term appears in the document.
LSI allows you to compare and cluster documents and find relationships between terms. I
criticized both my word-frequency example and the Naïve Bayes algorithm for not using the
relationships between words in a text sample; LSI *does* use word relationships. LSI determines
if word combinations often appear together in similar documents. Wikipedia has a good
article on LSI that I recommend as a reference: `http://en.wikipedia.org/wiki/Latent_`
`semantic_indexing`. In this section, you'll use the LSI library for both classification and summary generation.

■**Caution** LSI is covered by a U.S. patent, which might place limitations on how you use LSI in commercial
projects.

I'll wrap the LSI library in the LatentSemanticAnalysisClassifier class:

```ruby
require 'rubygems'
require 'classifier'

class LatentSemanticAnalysisClassifier
  def classify_plain_text text
    @lsi.classify(text)
  end
  def train file_and_topic_list
    topic_names = file_and_topic_list.collect {|x| x.last}
    @lsi = Classifier::LSI.new(:auto_rebuild => false)
    file_and_topic_list.each {|file, category|
      text = File.new(file).read
      @lsi.add_item(text, category)
    }
    @lsi.build_index
  end
  def get_summary text
    text.summary(2).gsub(' [...]', '.')
  end
end
```

Wrapping the LSI library in this class means you can have the same APIs that you used for my library from the "Using Word-count Statistics for Categorization" section and for the Bayesian classification library from the "Using a Bayesian Classifier for Categorization" section.

■**Note** You can use the same instance of LatentSemanticAnalysisClassifier by retraining it with different data. Calling the method classify_plain_text before calling the method train will cause a runtime error.

The method train uses a list of local text files with associated categories to use the LSI library APIs for training the classifier. Note that the LSI library can perform summarization without training data.

LSI takes longer than either the Bayesian method or my word-frequency code, and in my tests LSI has not provided as accurate results for document classification. The main reason for using LSI/LSA is the extra functionality for identifying similar document texts and discovering which words are frequently associated with one another. You can read through the examples and unit tests for the classifier gem if you want to experiment with LSI features that I am not using in this chapter.

When you load the classifier gem, the method String#summary is monkey-patched into the String class. This summary method takes an optional integer argument for the number of sentences to return as the summary (the default is to return ten sentences):

```
>> require 'classifier_lsa_using_classifier_gem'
=> true
>> test = LatentSemanticAnalysisClassifier.new
=> #<LatentSemanticAnalysisClassifier:0x1935584>
>> test.train([['wikipedia_text/computers.txt', "Computers"],
?>            ['wikipedia_text/economy.txt', "Economy"],
?>            ['wikipedia_text/health.txt', "Health"],
?>            ['wikipedia_text/software.txt', "Software"]])
>> require 'pp'
=> true
>> pp test.classify_plain_text("
Heart attacks and strokes kill too many people every year.")
"Computers"
=> nil
>> pp test.classify_plain_text("
Economic warfare rich versus the poor over international monetary fund.")
"Economy"
=> nil
>> pp test.classify_plain_text("My IBM PC broke so I bought an HP.")
"Computers"
=> nil
 >> text = File.new('wikipedia_text/health.txt').read[0..700]
=> "In 1948, the World Health Assembly defined health as \342\200\234a state of
complete physical, mental, and social well-being and not merely the absence of
disease or infirmity.\342\200\235 [1][2] This definition is still widely
referenced, but is often supplemented by other World Health Organization (WHO)
reports such as the Ottawa Charter for Health Promotion which in 1986 stated that
health is \342\200\234a resource for everyday life, not the objective of living.
Health is a positive concept emphasizing social and personal resources, as well as
physical capacities.\342\200\235\n\nClassification systems describe health. The
WHO\342\200\231s Family of International Classifications (WHO-FIC) is
composed of the International Classification "
>> test.classify_and_summarize_plain_text(text)
=> ["Economy", "\342\200\235 [1][2] This definition is still widely referenced,
but is often supplemented by other World Health Organization (WHO)
reports such as the Ottawa Charter for Health Promotion which in 1986
stated that health is \342\200\234a resource for everyday life, not the
objective of living. In 1948, the World Health Assembly defined health as
\342\200\234a state of complete physical, mental, and social well-being and
not merely the absence of disease or infirmity"]
>>
```

You notice that the classification of text fails in two of these examples because LSI requires a good deal of training-text data for each category. In this example, I was training with small data sets. In the next section, I'll show you how to use the Bayesian classifier and the LSI summary generator in the classifier gem.

Using Bayesian Classification and LSI Summarization

I am now going to create a new wrapper class that combines the document-classification functionality of the Bayesian library with the LSI functionality to generate summaries from text. This new class implements the public APIs for training, classification, and summarization that I've used throughout this chapter. This wrapper provides good results, but requires more processing resources (CPU time and memory). Here's the class definition for `ClassificationAndSummarization`, which combines code from the last two sections:

```ruby
require 'classifier'

class ClassificationAndSummarization
  def classify_plain_text text
    @bayes.classify(text)
  end
  def train file_and_topic_list
    topic_names = file_and_topic_list.collect {|x| x.last}
    @bayes = Classifier::Bayes.new(*topic_names)
    file_and_topic_list.each {|file, category|
      text = File.new(file).read
      @bayes.train(category, text)
    }
  end
  def get_summary text
    text.summary(2).gsub(' [...]', '.')
  end
  def classify_and_summarize_plain_text text
    [classify_plain_text(text), get_summary(text)]
  end
end
```

Using the same example text as you did in the last section (that is, the first 700 characters in the file `wikipedia_text/health.txt`), you now get good results. This test code snippet:

```ruby
test = ClassificationAndSummarization.new
test.train([['wikipedia_text/computers.txt', "Computers"],
            ['wikipedia_text/economy.txt', "Economy"],
            ['wikipedia_text/health.txt', "Health"],
            ['wikipedia_text/software.txt', "Software"]])
require 'pp'

text = File('wikipedia_text/health.txt').read[0..700]

pp test.classify_and_summarize_plain_text(text)
```

produces the following output:

```
["Health",
 "This definition is still widely referenced, but is often supplemented by other
```

World Health Organization (WHO) reports such as the Ottawa Charter for
Health Promotion which in 1986 stated that health is a resource for everyday life,
not the objective of living. In 1948, the World Health Assembly defined health as a
state of complete physical, mental, and social well-being and not
merely the absence of disease or infirmity"]

The important API in this class is the method classify_and_summarize_plain_text, which
I'll discuss in the "Automatically Generating Summaries" section when I present a combined
classifier and summary generator using word-count statistics. At the end of this chapter, I'll
integrate the word-frequency code into the class TextResource, but as an alternative you can
easily "drop in" the alternative class ClassificationAndSummarization and require the classifier gem.

You have seen two document classifiers: one using simple word-count statistics, and one
using the Naïve Bayes algorithm implemented in the classifier gem. Both of these offer results
of similar quality, so which one should you use? If you are also using the LSI library in the classifier gem for summarization, then it makes sense to use the classifier gem for classification
also. Plus, if you train an LSI classifier with large amounts of training data—using much more
memory and CPU time—then LSI will likely yield better classification results. That said, I usually use my own libraries that use word-count statistics for classification and summarization
because they require less memory and fewer CPU resources. At the end of this chapter, I'll
update the TextResource class from Chapters 1 and 2 with my code for word-count statistics.

The next two sections deal with entity extraction: extracting people's names and place
names from text. The example in the next section uses my own code, and the example in the
following section uses the Open Calais web service.

Extracting Entities from Text

For some applications, you might need to extract named entities such as people's names,
place names, and product names from text. The algorithm I use is straightforward if you have
word-list files for the types of names you want to identify in text. In the source code associated with this book (downloadable from the Apress web site), you'll find files in the src/part1/
data directory that contain large word lists for first names, last names, prefix names (Dr., Mrs.,
and so on), and place names. This directory also contains files with lists of words that have
negative and positive connotations; you'll use these files later in this chapter to calculate the
"sentiment" of text.

■**Tip** The Moby Word List project (http://icon.shef.ac.uk/Moby/ and http://infochimps.org/
dataset/word_list_mobywords_place_names) offers a set of word lists for use in statistical NLP programs. Another good source of linguistic data is the Penn Treebank Project (http://www.cis.upenn.
edu/~treebank/).

The entity-extraction algorithm has two parts. You first tokenize text into an array of
words and build a second array that has an empty array for each element. Using the word lists,

you fill these subarrays with tokens indicating possible entity-name types. On each line in the following example, a word in the input text is followed by an array of possible name-token types for the word:

```
President    [:last_name, :prefix_name]
George       [:first_name, :last_name]
W            []
Bush         [:last_name]
left         []
office       []
and          []
Barack       [:first_name]
Obama        [:last_name]
was          []
sworn        []
in           []
as           []
president    []
and          []
went         []
to           []
Florida      [:us_state, :first_name, :last_name]
with         []
his          []
family       []
to           []
stay         []
at           []
Disneyland   []
```

The second part of the algorithm applies a set of rules for removing possible token types for each word. You will see these rules in a Ruby script later in this section when I describe the implementation of the EntityExtraction class, so I will not list them here. After you apply rules for removing tokens, each word in the input text is tagged with the following token flags:

```
President    [:prefix_name]
George       [:first_name]
W            [:middle_initial]
Bush         [:last_name]
left         []
office       []
and          []
Barack       [:first_name]
Obama        [:last_name]
was          []
sworn        []
in           []
```

```
as              []
president       []
and             []
went            []
to               []
Florida         [:us_state]
with            []
his             []
family          []
to              []
stay            []
at              []
Disneyland      []
```

Here is a complete listing for the class EntityExtraction (I'll use the code-continuation character ➥ to break long lines). I start by defining constants for hash tables containing the name types that you need:

```
PLACE_NAMES = Hash.new
open('data/placenames.txt').readlines.each {|line|
  index = line.index(':')
  PLACE_NAMES[line[0...index].strip] = line[index+1..-1].strip
}
### note: possible values in PLACE_NAME hash:
#        ["city", "us_city", "country_capital", "country", "us_state"]
PLACE_STRING_TO_SYMBOLS = {'city' => :city, 'us_city' => :us_city,
                           'country_capital' => :country_capital,
                           'country' => :country,
                           'us_state' => :us_state}
PLACE_STRING_SYMBOLS = PLACE_STRING_TO_SYMBOLS.values

FIRST_NAMES = {}
open('data/firstnames.txt').readlines.each {|line|
  FIRST_NAMES[line.strip] = true
}
LAST_NAMES = {}
open('data/lastnames.txt').readlines.each {|line|
  LAST_NAMES[line.strip] = true
  puts "last: #{line}" if line.strip.index(' ')
}
PREFIX_NAMES = {}
open('data/prefixnames.txt').readlines.each {|line|
  PREFIX_NAMES[line.strip] = true
  puts "prefix: #{line}" if line.strip.index(' ')
}
```

You'll get a better understanding of this implementation of the EntityExtraction class later in this section, when you see printouts of sample text words followed by possible name-token types. The following code performs the calculations shown in these printouts:

```ruby
class EntityExtraction
  attr_getter :human_names
  attr_getter :place_names
  def initialize text = ''
    words = text.scan(/[a-zA-Z]+/)
    pp words
    word_flags = []
    words.each_with_index  {|word, i|
      word_flags[i] = []
      if PLACE_NAMES[word]
       word_flags[i] << PLACE_STRING_TO_SYMBOLS[PLACE_NAMES[word]]
      end
      word_flags[i] << :first_name  if FIRST_NAMES[word]
      word_flags[i] << :last_name   if LAST_NAMES[word]
      word_flags[i] << :prefix_name if PREFIX_NAMES[word]
    }
```

The following temporary debug printout produced the printout you'll see at the end of this section:

```ruby
    words.each_with_index {|word, i|
      puts "#{word}\t#{word_flags[i].join(' ')}"
    }
```

The remainder of the class definition is the implementation of the rules for removing possible name-token types from the array of tokens for each word in the input text:

```ruby
    # easier logic with two empty arrays at end of word flags:
    word_flags << [] << []

    # remove :last_name if also :first_name and
    # :last_name token nearby:
    word_flags.each_with_index  {|flags, i|
      if flags.index(:first_name) && flags.index(:last_name)
        if word_flags[i+1].include?(:last_name) ||
           word_flags[i+2].include?(:last_name)
           word_flags[i] -= [:last_name]
        end
      end
    }

    # look for middle initials in names:
    words.each_with_index {|word, i|
      if word.length == 1 && word >= 'A' && word <= 'Z'
        if word_flags[i-1].include?(:first_name) &&
           word_flags[i+1].include?(:last_name)
```

```
        word_flags[i] << :middle_initial if word_flags[i].empty?
      end
    end
  }

  # discard all but :prefix_name if followed by a name token:
  word_flags.each_with_index  {|flags, i|
    if flags.index(:prefix_name)
      word_flags[i] = [:prefix_name] ➥
            if human_name_symbol_in_list?(word_flags[i+1])
    end
  }

  # discard two last name tokens in a row if the preceding
  # token is not a name token:
  word_flags.each_with_index  {|flags, i|


    if i<word_flags.length-2 && ➥
       !human_name_symbol_in_list?(flags) && ➥
        word_flags[i+1].include?(:last_name) && ➥
       word_flags[i+2].include?(:last_name)
       word_flags[i+1] -= [:last_name]
    end
  }

  # discard singleton name flags (with no name flags on either side):
  word_flags.each_with_index  {|flags, i|
    if human_name_symbol_in_list?(flags)
      if !human_name_symbol_in_list?(word_flags[i+1]) &&
         !human_name_symbol_in_list?(word_flags[i-1])
        [:prefix_name, :first_name, :last_name].each {|name_symbol|
         word_flags[i] -= [name_symbol]
        }
      end
    end
  }
```

The following code uses the word-type tokens for each word in the input text to build arrays containing human and place names; you'll remove duplicate array elements later:

```
@human_names = []
human_name_buffer = []
@place_names = []
place_name_buffer = []
in_place_name = false
in_human_name = false
word_flags.each_with_index  {|flags, i|
```

```
      in_human_name = human_name_symbol_in_list?(flags)
      if in_human_name
        human_name_buffer << words[i]
      elsif !human_name_buffer.empty?
        @human_names << human_name_buffer.join(' ')
        human_name_buffer = []
      end
      in_place_name = place_name_symbol_in_list?(flags)
      if in_place_name
         place_name_buffer << words[i]
      elsif !place_name_buffer.empty?
        @place_names << place_name_buffer.join(' ')
        place_name_buffer = []
      end
    }
    @human_names.uniq! # remove duplicate elements
    @place_names.uniq!   # remove duplicate elements
  end
```

The remainder of the class implementation consists of private utility methods:

```
private

  def human_name_symbol_in_list? a_symbol_list
    a_symbol_list.each {|a_symbol|
      return true if [:prefix_name, :first_name, :middle_initial,
                      :last_name].index(a_symbol)
    }
    false
  end

  def place_name_symbol_in_list? a_symbol_list
    a_symbol_list.each {|a_symbol|
       return true if PLACE_STRING_SYMBOLS.include?(a_symbol)
    }
    false
  end
end
```

Finally, here is the output from running this example:

```
>> require 'entity_extraction'
=> true
>> ee = EntityExtraction.new('President George W. Bush left office and Barack Obama
was sworn in as president and went to Florida with his family to stay
at Disneyland.')
=> #<EntityExtraction:0x2e2a4f8 @human_names=["President George W Bush",
"Barack Obama"], @place_names=["Florida"]>
```

```
>> pp ee.human_names
["President George W Bush", "Barack Obama"]
=> nil
>> pp ee.place_names
["Florida"]
=> nil
```

The EntityExtraction class is simple and efficient. Best of all, this class produces good results when you have good word lists for different types of names. In the next section, I'll provide an alternative strategy for identifying entity names using the state-of-the-art Open Calais web service.

Performing Entity Extraction Using Open Calais

You can use the Open Calais web service free of charge for a limited number of web-service calls per day. If you want to use the code in this section, you will need to get a free developer's key at http://www.opencalais.com and set this key as the environment variable OPEN_CALAIS_KEY. Here's how to test that you have your key set:

```
>> ENV['OPEN_CALAIS_KEY']
=> "abc123-not-a-real-key-890xyz"
```

The Open Calais software was developed by ClearForest and purchased by Thomson Reuters, which made the service available for free (up to 20,000 web-service calls a day) and as a supported product. You can encode text from any source in a web-service call, and the returned payload contains detailed semantic information about the submitted text. The Open Calais web service returns Resource Description Framework (RDF) data encoded in XML. I'll hold off on a detailed discussion about RDF until Chapter 4, so for the example in this chapter I'll discard the RDF data returned from the web-service call. For the purposes of this section, I'll simply use a comment block that includes a list of named entities recognized in input text submitted to the Open Calais web service. Here's the comment block, reformatted for better readability:

```
<?xml version="1.0" encoding="utf-8"?>

<string xmlns="http://clearforest.com/">
<!--Use of the Calais Web Service is governed by the Terms of Service located at
http://www.opencalais.com. By using this service or the results of the service
you agree to these terms of service.-->
<!--Relations:
    Company: Pepsi
    Organization: White House
    Person: Barack Obama, Hillary Clinton
    ProvinceOrState: Texas-->
```

The rest of the code snippets in this section comprise the implementation for my client Ruby gem, gem `calais_client`. You will need to gem `install simplehttp` before running this example:

```
require 'simple_http'
require "rexml/document"
include REXML

module CalaisClient
  VERSION = '0.0.1'
  MY_KEY = ENV["OPEN_CALAIS_KEY"]
  raise(StandardError,"No Open Calais login key in ENV: '➥
OPEN_CALAIS_KEY'") if !MY_KEY
```

The constant `PARAMS` is a string containing the partial URL for a representational state transfer (REST) web-service call (see Chapter 13) to the Open Calais web service. The last part of this string was copied from Open Calais documentation:

```
  PARAMS = "&paramsXML=" + CGI.escape('<c:params xmlns:c=
"http://s.opencalais.com/1/pred/" xmlns:rdf="➥
http://www.w3.org/1999/02/22-rdf-syntax-ns#"><c:processingDirectives ➥
c:contentType="text/txt" c:outputFormat="xml/rdf"></c:processingDirectives>➥
<c:userDirectives c:allowDistribution="true" c:allowSearch=➥
"true" c:externalID="17cabs901" c:submitter="ABC">➥
</c:userDirectives><c:externalMetadata></c:externalMetadata></c:params>')
```

The class constructor for `OpenCalaisTaggedText` CGI-escapes input text so that it can be part of a URL and combines this with the `PARAMS` string and login information to make the web-service call:

```
  class OpenCalaisTaggedText
    def initialize text=""
      data = "licenseID=#{MY_KEY}&content=#{CGI.escape(text)}"
      http =  SimpleHttp.new(
        "http://api.opencalais.com/enlighten/calais.asmx/Enlighten")
      @response = CGI.unescapeHTML(http.post(data+PARAMS))
    end
    def get_tags
      h = {}
      index1 = @response.index('terms of service.-->')
      index1 = @response.index('<!--', index1)
      index2 = @response.index('-->', index1)
      # save the text between '<!--' and '-->':
      txt = @response[index1+4..index2-1]
      lines = txt.split("\n")
      lines.each {|line|
        index = line.index(":")
        h[line[0...index]] =
          line[index+1..-1].split(',').collect {|x| x.strip} if index
      }
```

```ruby
      # now, just keep the tags we want:
      ret = {}
      ["City", "Organization", "Country", "Company"].each {|ttype|
        ret[ttype] = h[ttype]
      }
      vals = []
      h["Person"].each {|p|
        vals << p if p.split.length > 1
      }
      ret["Person"] = vals
      ret["State"] = h["ProvinceOrState"]
      ret
    end
    def get_semantic_XML
      @response
    end
    def pp_semantic_XML
      Document.new(@response).write($stdout, 0)
    end
  end
end
```

The get_tags method parses the returned payload, ignoring the XML data and parsing what you need from the comment block. (In Chapters 4 and 11, I'll show you how to use the XML data from Open Calais web-service calls to populate a local RDF data store.) The get_semantic_XML method returns the XML (RDF) payload as a string. The pp_semantic_XML method parses the XML payload and "pretty-prints" it so you can easily read it (the returned payload is compressed and difficult to read).

For now, you'll use the named entities from the XML comment block and return to the RDF payload in Chapter 5. You can download the source code for my gem calais_client from the Apress web site, or you can install it using gem install opencalais_client. Here is an example using the client gem:

```ruby
>> require 'calais_client'
=> true
>> s = "Hillary Clinton and Barack Obama campaigned in Texas. Both want to live
in the White House. Pepsi sponsored both candidates."
=> "Hillary Clinton and Barack Obama campaigned in Texas. Both want to live in the
White House. Pepsi sponsored both candidates."
>> cc = CalaisClient::OpenCalaisTaggedText.new(s)
=> #<CalaisClient::OpenCalaisTaggedText:0x1a384cc @response= ... "
>> pp cc.get_tags
{"City"=>nil,
 "Organization"=>["White House"],
 "Person"=>["Barack Obama", "Hillary Clinton"],
 "Country"=>nil,
 "State"=>["Texas"],
```

```
  "Company"=>["Pepsi"]}
=> nil
```

There are three public methods in the class OpenCalaisTaggedText:

```
>> (cc.public_methods - Object.public_methods).sort
=> ["get_semantic_XML", "get_tags", "pp_semantic_XML"]
```

XML in these method names indicates that you will be using RDF serialized as XML data. You'll use the get_semantic_XML method in later chapters when you want to access the full RDF payload from the Open Calais web service. The method pp_semantic_XML is useful if you want to "pretty-print" the RDF XML payload for readability.

Automatically Generating Summaries

I will rewrite the class ClassifierWordCountStatistics in this section so it calculates text summaries in addition to performing its core responsibilities. The new class name will be ClassifierAndSummarization. This rewrite is for efficiency because classification and summarization share some of the same intermediate calculations. If you need only to classify text, then you can use the original class ClassifierWordCountStatistics.

■**Note** Here I'm presenting the ClassifierAndSummarization class, a combined classifier and summary generator using word-count statistics. This is not to be confused with the ClassificationAndSummarization class presented in the "Using Bayesian Classification and LSI Summarization" section, which is a wrapper class that combines the document-classification functionality of the Bayesian library with LSI functionality to generate text summaries.

In the algorithm I use for summarization, I first identify the categories automatically assigned to text and then identify which words in the text provided support for these categories. I then create an array containing the sentences in the original input text and assign a score to each sentence based on how many of the sentence's words contributed to classifying the original text. In the following code snippet, I am assuming that the String#each_char and sentence_boundaries methods are defined (I described these methods in Chapter 2). I show only the method classify_and_summarize_plain_text; otherwise, everything in this new class is identical to the original ClassifierWordCountStatistics class:

```
def classify_and_summarize_plain_text text
  word_stems = text.downcase.scan(/[a-z]+/)
  scores = Array.new(@category_names.length)
  @category_names.length.times {|i|
    scores[i] = score(@category_wc_hashes[i], word_stems)
  }
  best_index = scores.index(scores.max)
  best_hash = @category_wc_hashes[best_index]
```

```ruby
breaks = sentence_boundaries(text)
sentence_scores = Array.new(breaks.length)
breaks.length.times {|i| sentence_scores[i] = 0}
breaks.each_with_index {|sentence_break, i|
  tokens =
    text[sentence_break[0]..sentence_break[1]].scan(/[a-zA-Z]+/
  tokens.each {|token| sentence_scores[i] += best_hash[token]}
  sentence_scores[i] *= 100.0 / (1 + tokens.length)
}
score_cutoff = 0.8 * sentence_scores.max
summary = ''
sentence_scores.length.times {|i|
  if sentence_scores[i] >= score_cutoff
    summary << text[breaks[i][0]..breaks[i][1]] << ' '
  end
}
[@category_names[best_index], summary]
end
```

I normalized the score for each sentence by the number of words in the sentence and scaled the scores by 100 to make them more readable. For the final summary, I kept all sentences that had a score of at least 80 percent of the maximum sentence score. Here is an example showing how to use this class:

```ruby
>> require 'classification_and_summarization'
Running Ruby 1.8.x: define each_char
=> true
>> test = ClassifierAndSummarization.new
=> #<ClassifierAndSummarization:0x1a5eeec    ... >
>> test.train([['wikipedia_text/computers.txt', "Computers"],
?>             ['wikipedia_text/economy.txt', "Economy"],
?>             ['wikipedia_text/health.txt', "Health"],
?>             ['wikipedia_text/software.txt', "Software"]])
=> [["wikipedia_text/computers.txt", "Computers"], ["wikipedia_text/economy.txt",
"Economy"], ["wikipedia_text/health.txt", "Health"],
["wikipedia_text/software.txt", "Software"]]
>> pp test.classify_and_summarize_plain_text("Doctors advise exercise to improve
heart health. Exercise can be can be as simple as walking 25 minutes per day.
A low fat diet is also known to improve heart health. A diet of fast food
is not recommended.")
["Health",
 "Doctors advise exercise to improve heart health. A low fat diet is also known
to improve heart health."]
```

While I sometimes use the classifier Ruby gem, I usually use my simpler and much less computationally expensive word frequency–based library. I suggest that you try both approaches in your application to see what works best for you.

Determining the "Sentiment" of Text

The goal of this section is to write a Ruby class for determining the sentiment of text. I have created two text data files: one containing words with positive connotations, and the other containing words with negative connotations. I reopen the class `ClassifierWordCountStatistics` and monkey-patch it to add the method `get_sentiment`. The new class `SentimentOfText` contains an instance of `ClassifierWordCountStatistics` and uses the monkey-patched method:

```ruby
require 'classifier_word_count_statistics'

class ClassifierWordCountStatistics
  def get_sentiment text
    word_stems = text.scan(/[a-zA-Z]+/)
    scores = []
    2.times {|i|
      scores[i] = score(@category_wc_hashes[i], word_stems)
    }
    scores[0] - scores[1]
  end
end


class SentimentOfText
  def initialize
    @classifier = ClassifierWordCountStatistics.new
    @classifier.train([['data/positive.txt', "Positive"],
                       ['data/negative.txt', "Negative"]])
  end
  def get_sentiment text
    @classifier.classify_plain_text(text)[0]
  end
end
```

This strategy of using a relative count of "positive" and "negative" words works fairly well. I am not going to improve this strategy, but it is worth mentioning two difficult-to-implement improvements. The first improvement is to consider word pairs that have a strong positive or negative tone when both words appear together. Using word-pair statistics would require creating word pair–count hashes for all adjacent word pairs in the training text, which in turn would require a large increase in memory and CPU resources. The second improvement is to try to determine if positive or negative terms are negated in context. For example, in the sentence "Mr. Smith is not a bad man," the word "bad" is negated and should be ignored.

For the purposes of developing Web 3.0 applications, you should find the Ruby class `SentimentOfText` adequate for rating the sentiment of blog entries, web pages, and so on. Here is an example use of this class:

```
>> require 'sentiment-of-text'
=> true
>> st = SentimentOfText.new
=> #<SentimentOfText:0x1a3f844    ...  >
```

```
>> pp st.get_sentiment("the boy kicked the dog")
-0.11648223645894
 => nil
>> pp st.get_sentiment("the boy greeted the dog")
0.14367816091954
=> nil
```

This algorithm for calculating the sentiment of text sometimes produces wrong results, but it's adequate for roughly gauging the positive or negative sentiment of text. In a real application, you would want to use a very large set of training documents (that is, documents that are rated as "negative" or "positive" and used to build the negative and positive word-count hash tables).

An application where you might want to use the class `SentimentOfText` would be a system that searches for a company's products on the Web and keeps a running total of positive- or negative-sentiment web pages mentioning the products.

Clustering Text Documents

A cluster is a set of similar documents. For example, if you had a set of PDF files that covered either health or politics, then you would expect a clustering algorithm to separate the files into two groups: health files in one cluster and politics files in the other cluster. One approach for clustering documents is to use assigned categories (or classifications) by which documents contain references to the same people, places, products, and so on. Here is an application idea: a search engine might identify a set of documents using a simple keyword search, and each search result can show named entities with links to other documents containing the same names.

You can also cluster small numbers of documents by calculating the intersection of common words. This produces accurate results, but unfortunately scales as $O(n^2)$ where n is the number of documents, so you need to perform a calculation of each document pair. A more efficient algorithm for clustering large numbers of documents is *K-means clustering*.

K-means Document Clustering

Surendra Singhi has written a Ruby gem clusterer that supports K-means clustering. You can install it using `gem install clusterer`, which implements K-means and hierarchical clustering. This library has not been kept up-to-date since 2007 and it is not well-documented, but it contains some useful code. This gem also hooks into the Ruby on Rails `ActiveRecord` class framework to add an `acts_as_clusterable` behavior, which Rails developers should find useful.

■**Tip** Wikipedia has an article on K-means clustering: http://en.wikipedia.org/wiki/K-means.

I am not going to integrate clustering functionality into my `TextResource` class, so it suffices here to show a short example using this gem:

```
>> text1 = File.new('wikipedia_text/computers.txt').read
=> "A computer is a machine..."
>> text2 = File.new('wikipedia_text/economy.txt').read
=> "\nEconomy\nFrom Wikipedia, the ..."
>> text3 = File.new('wikipedia_text/health.txt').read
=> "In 1948, the World Health Assembly ..."
>> text4 = File.new('wikipedia_text/software.txt').read
=> "Software is a general term ..."
>> doc1 = Clusterer::Document.new(text1)
=> {"png"=>1, "earlier"=>2,
>> doc2 = Clusterer::Document.new(text2)
=> {"earlier"=>1, "parasit"=>1,
>> doc3 = Clusterer::Document.new(text3)
=> {"multidisciplinari"=>1, "behavior"=>2,
>> doc4 = Clusterer::Document.new(text4)
=> {"w3c"=>1, "whole"=>1, "lubarski"=>1,
>> cluster = Clusterer::Clustering.cluster(:kmeans,[doc1, doc2, doc3, doc4],
:no_of_clusters => 3)
Iteration ....0
Iteration ....1
=> [#<Clusterer::Cluster:0x2249e90 @documents=[{    .... .....
```

There are three clusters in the result. I would have expected four clusters, but the train-ing text for the Computers and Software categories were combined into a single cluster. The following example prints out the internal document IDs, which provide the only way to deter-mine which input documents are in each cluster:

```
>> cluster[0].documents.each {|doc| puts doc.object_id} ; :ok
 18078930
17997460
 => :ok
>> cluster[1].documents.each {|doc| puts doc.object_id} ; :ok
18015210
=> :ok
>> cluster[2].documents.each {|doc| puts doc.object_id} ; :ok
18040440
=> :ok
```

This output shows that the first cluster contains the articles on software and computers, the second cluster contains the document on the economy, and the third cluster contains the article on health.

Clustering Documents with Word-Use Intersections

You can compare two documents for similarity by finding the number of words that appear in both documents and scaling this by the number of word occurrences divided by the number of words in both documents. Documents with high similarity rankings belong in the same document cluster. This process can be generalized, but the runtime performance is O(n^2) where n is the number of documents.

Document clustering through word-use intersections gives good results, but does not scale well for clustering large numbers of documents. So for large document collections, you're better off using K-means clustering. For smaller document sets, I like to use word-use intersections because this technique yields good results. The following code shows the file word-count-similarity.rb, which calculates a similarity measure of two texts by calculating the word intersection of two text samples. Then it normalizes the result by dividing by the sum of the number of words in the two text samples:

```ruby
def word_use_simularity text1, text2
  tokens1 = text1.scan(/[a-zA-Z]+/)
  tokens2 = text2.scan(/[a-zA-Z]+/)
```

The following line of code produces an array of word tokens that appear in both the tokens1 and tokens2 arrays:

```ruby
  common_tokens = tokens1 & tokens2
  common_tokens.length.to_f / (tokens1.length + tokens2.length)
end
```

The same source file also contains a longer function that takes any number of strings and calculates similar clusters; clusters are identified by arrays containing argument indices:

```ruby
def cluster_strings *text_strings
  @texts = []
  text_strings.each {|text|
    tokens = text.downcase.scan(/[a-z]+/)
    @texts << tokens
  }
  # create a two-dimensional array [doc index][doc_index]:
  size = @texts.length
  similarity_matrix = Array.new(size)
  similarity_matrix.map! {
    temp = Array.new(size)
    temp.size.times {|i| temp[i] = 0}
    temp
  }
    size.times {|i|
    size.times {|j|
      common_tokens = @texts[i] & @texts[j]
      similarity_matrix[i][j] = ➥
          2.0 * common_tokens.length.to_f / ➥
          (@texts[i].length + @texts[j].length)
    }
  }
  # calculate possible clusters:
  cluster = []
  size.times {|i|
```

```
      similar = []
      size.times {|j|
        similar << j if j > i && similarity_matrix[i][j] > 0.1
      }
      cluster << (similar << i).sort if similar.length > 0
    }
    # remove redundant clusters:
    cluster.size.times {|i|
      cluster.size.times {|j|
        if cluster[j].length < cluster[i].length
          cluster[j] = [] if (cluster[j] & cluster[i]) == cluster[j]
        end
      }
    }
    result = []
    cluster.each {|c| result << c if c.length > 1}
    result
  end
```

Now I'll show you an example using both word_use_similarity and cluster_strings methods. If both arguments to word_use_similarity contain the same words, then this method returns its maximum possible value of 0.5. If there are no common words, then word_use_similarity returns zero. The method cluster_strings takes any number of string arguments and returns clusters containing the argument indices:

```
 >> require 'word-count-similarity'
=> true
>> s1="Software products"
=> "Software products"
>> s2="Hardware products"
=> "Hardware products"
>> s3="misc words matching nothing"
=> "misc words matching nothing"
>> s4="Software and Hardware products"
=> "Software and Hardware products"
>> s5="misc stuff"
=> "misc stuff"
>> word_use_similarity(s1, s2)
=> 0.25
>> word_use_similarity(s1, s3)
=> 0.0
>> word_use_similarity(s2, s3)
=> 0.0
>> word_use_similarity(s3, s5)
=> 0.166666666666667
>> cluster_strings(s1, s2, s3, s4, s5)
=> [[0, 1, 3], [2, 4]]
```

In the preceding example, the method `cluster_strings` is called with five arguments: the first calculated cluster contains the first, second, and fourth arguments (indices 0, 1, 3), and the second cluster contains the third and fifth arguments (indices 2 and 4).

In the next section, I will pull together examples in this chapter to augment the class `TextResource` that you saw in Chapters 1 and 2.

Combining the TextResource Class with NLP Code

I will now update the `TextResource` base class to use the NLP tools developed in this chapter. There are several tools to choose from: I showed you three options for classifying text and two options for determining named entities in text. The modified version of `TextResource` will use my word-count statistics approach for classification and my pure Ruby library for entity-name extraction from text.

You can find the source code for the `TextResource` class with the example snippets from Chapters 1 and 2 in `src/part1/text-resource_chapter2.rb` (downloadable from the Apress web site). The `text-resource.rb` file contains everything you need for:

- Reading all file formats covered in Chapter 1

- Performing text cleanup and sentence segmentation covered in Chapter 2 (spelling-correction examples are not included)

- Classifying documents based on word frequency, covered in this chapter

- Summarizing documents, covered in this chapter

- Extracting entity names (human and place names), covered in this chapter

- Calculating a sentiment rating, covered in this chapter

The file `text-resource.rb` is large by Ruby standards—about 400 lines of code—because it contains the base class `TextResource` and all its derived classes. The directory `src/part1` also offers all the snippets as standalone scripts if you want to use some of the material from the first three chapters in your own projects. When you run the examples in `src/part1`, you also will need the data files in these subdirectories:

- `src/part1/wikipedia_text`: Contains training-text files for categories "Computers," "Economy," "Health," and "Software"

- `src/part1/data`: Contains word-list files for first names, last names, name prefixes, place names, negative sentiments, and positive sentiments

The functionality added to class `TextResource` appears largely in the definition of the following method that was stubbed out in Chapter 1:

```ruby
def process_text_semantics text
  cs = ClassifierAndSummarization.new
  cs.train([['wikipedia_text/computers.txt', "Computers"],
            ['wikipedia_text/economy.txt', "Economy"],
            ['wikipedia_text/health.txt', "Health"],
            ['wikipedia_text/software.txt', "Software"]])
  results = cs.classify_and_summarize_plain_text(@plain_text)
```

```
  @categories = results[0]
  @summary = results[1]
  @summary = @title + ". " + @summary if @title.length > 1
  @sentence_boundaries = get_sentence_boundaries(@plain_text)
  ee = EntityExtraction.new(@plain_text)
  @human_names = ee.human_names
  @place_names = ee.place_names
  st = SentimentOfText.new
  @sentiment_rating = st.get_sentiment(@plain_text)
end
```

It is my intention that you will modify process_text_semantics by substituting your own training files for the categories you need and by building a classifier and reusing it. Remember that automatic summarization depends on automatic classification to work properly. By making these changes and copying the code snippets from this chapter to the source file text-resource.rb, you should be able to use most of the following attributes for documents of any type:

```
class TextResource
  attr_accessor :source_uri
  attr_accessor :plain_text
  attr_accessor :title
  attr_accessor :headings_1
  attr_accessor :headings_2
  attr_accessor :headings_3
  attr_accessor :sentence_boundaries
  attr_accessor :categories
  attr_accessor :place_names
  attr_accessor :human_names
  attr_accessor :summary
  attr_accessor :sentiment_rating # positive number implies positive sentiment
```

For some document types, the headings attributes do not get set. The example code for the first three chapters is available in two forms: as individual Ruby source files, and as an aggregation in the large text-resource.rb source file. If you need specific code in your application, use the shorter script files.

Now I'll show you how to write a simple application using the class TextResource and the subclass PlainTextResource. Namely, I'll write a function that takes a list of file paths to text files and calculates the list of human names that appear in every file (that is, the intersection of names that appear in all of the specified files). I'll list this function and then test it using two text files created from about one-third of the text from two Wikipedia articles on Hillary Clinton and Barack Obama. Here is the example code:

```
require 'text-resource'

def find_common_names *file_paths
  names_lists =
    file_paths.map{|file_path|
      PlainTextResource.new(file_path).human_names
```

```
      }
   common_names = names_lists.pop
   names_lists.each {|nlist| common_names = common_names & nlist}
   common_names
end
```

If you are running irb in the src/part1 directory, your output will look like this:

```
>> find_common_names("test_data/wikipedia_Hillary Rodham Clinton.txt",
"test_data/wikipedia_Barack_Obama.txt")
=> ["Barack Obama", "John McCain"]
```

This function would be much more efficient if you would just use the class EntityExtraction because the PlainTextResource constructor performs full semantic processing on input text. Still, unless performance is a real issue, I sometimes prefer having "Swiss Army knife" utilities that perform a wide range of services.

Here is a more efficient version of this sample application that uses only the EntityExtraction class. This example has additional functionality—it finds the intersection of all place names appearing in all of the input text files:

```
require 'entity_extraction'

def find_common_names *file_paths
   names_texts = file_paths.map{|file_path| File.new(file_path).read}
   extractors = names_texts.map{|file_path|
                  EntityExtraction.new(file_path)
                }
   # human names:
   names_lists = extractors.map{|extractor| extractor.human_names}
   common_names = names_lists.pop
   names_lists.each {|nlist| common_names = common_names & nlist}
   # place names:
   places_lists = extractors.map{|extractor| extractor.place_names}
   common_places = places_lists.pop
   places_lists.each {|nlist| common_places = common_places & nlist}
   # return both human and place names appearing in all input texts:
   [common_names, common_places]
end
```

If you are running irb in the src/part1 directory, you can try this example:

```
>> find_common_names("test_data/wikipedia_Hillary Rodham Clinton.txt",
"test_data/wikipedia_Barack_Obama.txt")
=> [["Barack Obama", "John McCain"],
      ["Illinois", "Columbia", "Chicago", "Iraq", "Washington", "World"]]
```

■**Warning** You might make the mistake of trying to instantiate instances of the base class `TextResource` that perform no processing of input text. Instead, make sure that you use a subclass such as `PlainTextResource`, `HtmlXhtmlResource`, `OpenDocumentResource`, or `RssResource`.

Wrapup

In Chapter 1, you learned how to develop a common API for using text from different sources: different file formats, HTML, RSS feeds, and Atom feeds. In Chapter 2, you learned two useful techniques: cleaning up text and performing automatic spelling correction.

I showed you several useful statistical NLP techniques in this chapter. You learned how to categorize text automatically using different methodologies, extract entity names from text, generate summaries, determine whether text sentiment is positive or negative, and cluster text documents through different means. Plus, I showed you how to augment the base `TextResource` class with the NLP code that I presented throughout the chapter.

In addition to these NLP techniques, I would like you to take away something else: extracting semantic information from text can be useful in a wide variety of applications. You will find, however, that you usually need to experiment with sample input texts for your domain and adjust algorithms and code based on your own experiments. Canned libraries offer a good starting point, but please be prepared to spend time tailoring libraries like the ones we used in this chapter to your own needs. In later chapters, I'll show you how to store semantic information in data stores and use it to access information as an alternative to using simple search techniques. In the last few decades, some of my more enjoyable work has involved the extraction of information from text. I'll continue with the theme of finding and using information in Part 3.

In Part 2, you'll examine what I believe will be a core and important technology for Web 3.0 applications: the Semantic Web. You will see how you can use standard data formats such as RDF, RDF Schema (RDFS), and OWL to encode information for use by software agents. I believe that Web 3.0 applications will be characterized by their ability to provide data to both human readers and software agents. You'll also explore how to efficiently produce interfaces for both human and software clients in both Parts 2 and 4.

PART 2

■■■

The Semantic Web

In this first chapter of Part 2, I'll provide an introduction to Semantic Web data standards and the SPARQL query language, followed by some example code using SPARQL queries to RDF data stores on the Web. You will learn how to set up and use your own RDF data stores in Chapter 5. In Chapter 6, I'll go into more detail using SPARQL and reasoning to integrate and use Semantic Web data sources. Chapter 7 will show you how to build a SPARQL endpoint web portal.

CHAPTER 4

Using RDF and RDFS Data Formats

As I mentioned at the end of the previous chapter, I believe that a key technology for Web 3.0 will be the Semantic Web. The Semantic Web concept, which was originally introduced in a *Scientific American* paper,[1] embodies the idea that web applications can provide useful data to both human readers and software agents. The publication of Semantic Web content along with the development of related tools has been and continues to be a slow process. Here's an analogy: one fax machine is useless; you need a critical mass of installed fax machines before you get the "network-effect" benefit. The same thing holds for the Semantic Web: tools get written as more content is published in software-readable forms, and more content is produced as more tools get written.

The Semantic Web involves more than simply providing data and exchanging information; the goal is to share knowledge. The study of "knowledge representation" is key to the field of artificial intelligence (AI), but it is another technology that will likely never be finished. The process of developing knowledge-representation technologies has influenced the design of data formats and logic systems that work with these formats.

This chapter serves as a light introduction to the Semantic Web using Resource Description Framework (RDF) and RDF Schema (RDFS). I will not cover logic in any detail and I'll only slightly cover the most expressive Semantic Web modeling language, Web Ontology Language (OWL). OWL is more difficult to use than RDF/RDFS, and while RDF/RDFS is more lightweight, it suffices for a wide range of applications. My goal is to get you started on the journey of implementing Semantic Web technologies in your Web 3.0 applications.

Note You'll see the term "ontology" used throughout this chapter. An ontology is a definition of a set of concepts and the relationships between those concepts, which you can use to model a knowledge domain.

---

1. Tim Berners-Lee, James Hendler, and Ora Lassila, "The Semantic Web," http://www.sciam.com/article.cfm?id=the-semantic-web, May 2001.

You'll be concentrating on practical Semantic Web technologies. Figure 4-1 shows a subset of data formats that you will be using.

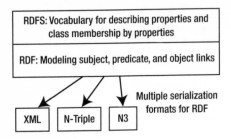

Figure 4-1. *The subset of Semantic Web data formats you'll be using*

Understanding RDF

RDF was designed as a universal data format for storing data as *triples*. A triple consists of data for:

- *Subject*: A Uniform Resource Identifier (URI) specifying the subject of the triple
- *Predicate*: A URI specifying a predicate, property, or relationship between a subject and an object
- *Object*: A URI or literal specifying the object of the triple

---

■**Note** My goal is to provide you with a gentle introduction to a difficult subject by giving you enough information to get started on your own projects. After working through this chapter and getting some hands-on experience with using RDF data in Chapter 5, I suggest that you revisit RDF more formally with the tutorial at http://www.w3.org/TR/REC-rdf-syntax.

URIs used to define triples represent resources on the Web; for example, the URI of a web page is its URL. Triple subjects and predicates are almost always specified as URIs, but objects might sometimes be specified as literal values. Because URIs represent unique identifiers for resources, the advantage of using URIs is that different data sources can use the same URI for the same "thing" in the physical world or the same abstract concept.

An example will show how important it is to use fixed URIs rather than literal values. Suppose that publishers of publicly available RDF data stores used string literals "Bill Clinton" and "Barack Obama." References to "Barack Obama" could very well be interpreted as representations of a person who is the 44th President of the United States because he has a fairly unique name. The situation is less clear with "Bill Clinton" because many other people share that name with the 42nd President of the United States. If a public RDF store defines data for the 42nd President and defines a unique URI to represent this person, then other information publishers can use this same URI and thus prevent ambiguity. What if different public RDF data stores define their own URIs for an individual? You'll see in the "Understanding RDFS" section how to use the RDFS language to make statements to declare the equivalency of different URIs in your applications.

There are common RDF schemas that define standard URIs for many predicates. An early and widely used project is the Dublin Core Metadata Initiative. I'll frequently use the Dublin Core definitions, which feature a namespace abbreviation `dc:` for RDF triple predicates such as `dc:title` and `dc:author`. In examples, I will often use the namespace abbreviation `kb:` for my knowledgebooks.com web site as a source of unique URIs.

■Tip There are other standards for URIs of subjects and objects, such as the Simple Knowledge Organization System (`http://www.w3.org/2004/02/skos/`), the OpenCyc world-knowledge ontology expressed in OWL (`http://opencyc.org/`), and the GeoNames ontology (`http://www.geonames.org/ontology/`). Whenever possible, try to reuse standard URIs for specific people, places, and concepts.

One strategy is employing a service such as Open Calais and using its URIs for subjects and objects. (Read more about Open Calais in Chapter 3.) For example, if I submit text containing "President Bill Clinton" to Open Calais, the returned RDF (serialized as XML) contains this (shortened for readability):

```
<rdf:RDF xmlns:c='http://s.opencalais.com/1/pred/'
        xmlns:rdf='http://www.w3.org/1999/02/22-rdf-syntax-ns#'>
 <rdf:Description rdf:about='http://d.opencalais.com/ph-1/f4710a'>
  <rdf:type rdf:resource='http://opencalais.com/1/type/e/Person'/>
      <c:name>Bill Clinton</c:name>
 </rdf:Description>
</rdf:RDF>
```

I expect that you, dear reader, find this as difficult to read and interpret as I did when I started learning how to use RDF and tried reading XML representation and notation of RDF data. XML serialization is useful for communication between software systems, but the XML notation is counterintuitive for expressing triple data. From now on, I'll use both N-Triple and N3 formats, and you will soon see some simple examples that are also expressed as graphs. You'll find both N-Triple and N3 formats easy to write, read, and understand. (Read more about these formats at `http://www.w3.org/2001/sw/RDFCore/ntriples/`.) N-Triple data is expressed using this format:

```
subject1 predicate1 object1 .
subject1 predicate2 object2 .
subject2
   predicate3
   object3 .
```

Notice that triples can be split over multiple lines. A triple in N-Triple format has three values and is terminated with a space character followed by a period. The N3 format is similar to N-Triple, but it allows triples with the same subject to be expressed more concisely:

```
subject1 predicate1 object1 ;
        predicate2 object2 .
subject2 predicate3 object3 .
```

N3 also allows you to collapse triples that share both subject and predicate:

```
subject1 predicate1 object1 ;
                    object2 ;
                    object3 .
```

Plus, N3 supports a notation for defining short namespace prefixes such as dc: for Dublin Core. The following listing shows an equivalent N-Triple representation of the preceding XML example, parsed by the Redland RDF parser (see Chapter 5). Note that URIs are enclosed as well-formed XML elements:

```
<http://d.opencalais.com/pershash-1/f6ef4710a' />
    <http://s.opencalais.com/1/pred/name />
    "Bill Clinton" .

<http://d.opencalais.com/pershash-1/f6ef4710a' />
    <http://www.w3.org/1999/02/22-rdf-syntax-ns#type />
    <http://s.opencalais.com/1/type/em/e/Person />  .
```

The subject in each of these triples is a unique URI that the Open Calais system assigns to the concept of "President Bill Clinton." The predicate value in the first triple is a URI for Open Calais's definition of a name. The object of the first triple is the string literal "Bill Clinton." The predicate for the second triple is the World Wide Web Consortium's definition for an RDF data type. The object of the second triple is a URI specifying Open Calais's definition of a "person type."

The N-Triple format is fine to use, and I especially like it when I am writing utility scripts to convert data to RDF. But I find the N3 format easier to read, so I sometimes convert data in N-Triple format to N3 format if I need to spend much time reading it. Here are the same two triples representing Bill Clinton in N3 format:

```
@prefix oc_type:    <http://s.opencalais.com/1/type/em/e#>
@prefix oc_pred:    <http://s.opencalais.com/1/pred#>
@prefix rdf:        <http://www.w3.org/1999/02/22-rdf-syntax-ns#">

<http://d.opencalais.com/pershash-1/f6ef4710a' />
                    oc_pred:name "Bill Clinton" ;
                    rdf:type oc_type:Person .
```

These N-Triple and N3 representations mean the same thing, and are also identical to the graph notation in Figure 4-2. The first three lines define prefix abbreviations for three XML namespaces. The last two lines represent two N3 triples that share the same subject. N3 data streams will typically be large, so the overhead of specifying namespaces is relatively small in terms of storage requirements. This type of representation makes triples much easier to read and understand.

Note You will see many examples of storing and accessing RDF—for now, I just want you to understand the syntax. You will become more comfortable with the N-Triple and N3 data formats when you use them in the "Understanding RDFS" section.

RDF data is often displayed as a graph, as seen in Figure 4-2. Triple subjects and objects are shown as nodes in the graph, and predicates are shown as directed arcs between nodes.

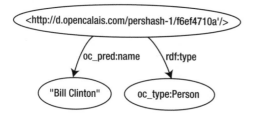

Figure 4-2. *Visualizing the last N-Triple and N3 RDF examples as a directed graph*

If you have not previously used RDF, you are probably questioning the wisdom of limiting a storage model to three values. To understand this limitation and how to work around it, consider the relational database table shown in Table 4-1.

Table 4-1. *A Database Table with Six Columns for Representing a Customer with One Row of Data*

customer_id	customer_name	email	zip_code	current_balance	invoice_date
1	Acme	admin@acme.com	85551	1510.00	2009/05/01

This table has six columns, so it will take six RDF triples to represent the data in one row. The common subject for all six triples can simply be the unique customer ID. Figure 4-3 shows an RDF graph representation of the database table in Table 4-1. I am using `rdf:` as an abbreviation for the RDF namespace and `test:` as an example test namespace.

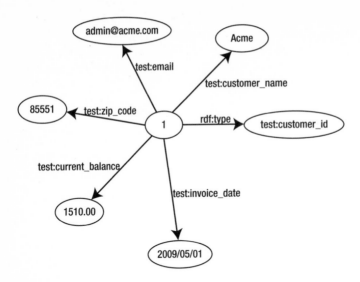

Figure 4-3. *A graph representation of the database example using Graphviz*

You might be wondering how I generated Figures 4-2 and 4-3. I used the Graphviz utility that I'll also use for visualizing RDF data in Chapter 7. (Find more details at http://www. graphviz.org/.) The Graphviz "data code" for a small part of Figure 4-3 is:

```
digraph database_sample {
    i1 [label="1"];
    n1 [label="Acme"]
    i1->n1 [label="test:customer_name"];
}
```

This notation creates two named nodes (i1 and n1) and an arc with the label test: customer_name that connects them. I generated this Graphviz data by hand, but in Chapter 7 I'll show you how to generate Graphviz visualization data with Ruby scripts using the graphviz Ruby gem. When you start using SPARQL Protocol and RDF Query Language (SPARQL), you'll find it useful to show graph representations of query results.

A drawback to using RDF that takes some time and effort to work around is the necessity to use multiple triples with the same subject to represent what you might prefer to handle as a single row in a relational database table or as a single object with multiple attributes in a Ruby object. There are six RDF triples represented in Figure 4-3.

The primary advantage of using RDF is that you can easily extend data models. If you want to add a column in a database table, for example, you simply need to change the schema and migrate the table. Furthermore, if you want to add a new attribute for only a few customers, you could easily add an additional table with the customer ID as a key. This process is simpler when you use RDF data stores: you add a triple with the new attribute name represented as a predicate. Using our example for a customer with ID = 1, you might add a triple with the subject equal to a URI representing this customer, a predicate of type test:second_email, and a literal string value for the secondary e-mail address.

You can use blank nodes (*bNodes*) to represent resources that do not have or do not need a name. You can use bNodes to represent subject and object parts of a triple or to group data that you might, for example, want to reify. Reification means to make a statement in a language using the same language. An example of using reification is to express a belief that some statement is true. Using a blank node _bn1, you might express:

```
_bn1 rdf:type rdf:Statement .
_bn1 rdf:subject <URI for customer 1> .
_bn1 rdf:predicate test:email .
_bn1 rdf:object "admin@acme.com" .
 test:Mark test:believes _bn1 .
```

The meaning of these five triples is that "Mark believes that the e-mail address for customer 1 is admin@acme.com."

RDF is a notation to associate predicates (or properties) to resources. In the next section, you'll examine the richer modeling language of RDFS.

Understanding RDFS

RDF provides limited semantics to data through the use of predicates (properties) with meaning that is assumed to be known. Now I'll show you a richer modeling language: RDF Schema (RDFS). You can use RDFS to define new predicates and to help merge data from different sources. Throughout this discussion, I will use the terms "property" and "predicate" interchangeably. These two terms both refer to the second value in RDF triples.

You'll use RDFS to define new classes and properties. To start an example, define a class Organization using N-Triple format:

```
<http://knowledgebooks.com/test#Organization />
    <http://www.w3.org/2000/01/rdf-schema#label />
    "Organization" .

<http://knowledgebooks.com/test#Organization />
    <http://www.w3.org/1999/02/22-rdf-syntax-ns#type />
    <http://www.w3.org/2000/01/rdf-schema#Class /> .
```

I started this example using N-Triple format because I want you to visualize triples in this form—and definitely not as an XML serialization. I also want you to be comfortable with the N-Triple notation. But now I'll switch to N3 for readability:

```
@prefix rdf:      <http://www.w3.org/1999/02/22-rdf-syntax-ns#">
@prefix rdfs:     <http://www.w3.org/2000/01/rdf-schema#>
@prefix kb:       <http://knowledgebooks.com/test#>
@prefix xml:      <http://www.w3.org/2001/XMLSchema#string>
kb:Organization rdfs:label "Organization" ;
              rdf:type rdfs:Class .
```

Assume the same three prefix abbreviations will apply as I continue this example. I will define a subclass of `Organization` called `Business`, and then a subclass of `Business` called `Customer`:

```
kb:Business rdfs:label "Business" ;
            rdfs:subClassOf kb:Organization ;
            rdf:type rdfs:Class .

kb:Customer rdfs:label "Customer" ;
            rdfs:subClassOf kb:Business ;
            rdf:type rdfs:Class .
```

The next example defines the new property (or predicate) `organizationName` with a range of type `string` and a domain of type `Organization`:

```
kb:organizationName rdfs:range xml:string ;
                    rdfs:domain kb:Organization ;
                    rdf:type rdf:Property .
```

This is a contrived example, but it shows you the necessary basics to model classes and properties for your application. Notice that class names are capitalized and that property names start with a lowercase letter. I show this model in Figure 4-4. The N3 format has a useful abbreviation: you can substitute the word "a" for `rdf:type`. For example, the following two statements are equivalent:

```
kb:Business rdf:type rdfs:Class .
kb:Business a rdfs:Class .
```

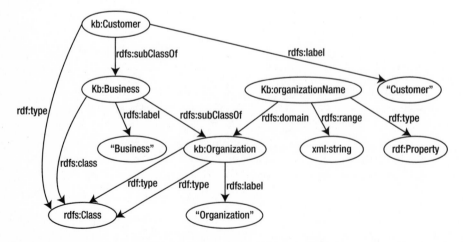

Figure 4-4. *RDFS example defining three classes and one property*

Not all RDF statements need to be stated explicitly. For example, RDFS also supports `rdfs:subProperty`, which is an instance of the class `rdf:Property`. If you make statements like the following that declare a subproperty and a subproperty value, then the value also holds for the parent property:

```
kb:companyName rdfs:subProperty kb:organizationName .
kb:testCompany1 kb:companyName "Acme Software Services" .
```

If you make a query (which you'll do formally in the "Creating SPARQL Queries" section) such as "show all X where X has a `kb:organizationName` of 'Acme Software Services'," then you'll get a match X == `kb:testCompany1`.

■**Caution** You'll examine several RDF data stores in Chapter 5. Different data stores provide different levels of support for inference. For example, I often use the Redland RDF libraries to manage and query RDF data, but Redland has limited RDFS-inferencing support. The AllegroGraph RDF data store supports RDFS inferencing and limited OWL inferencing. Sesame supports most of OWL, as does the Pellet reasoner used in the Protégé ontology editor.

Understanding OWL

I am not going to cover OWL in any detail, but I do want to let you know how OWL fits in as a Semantic Web modeling language. OWL is more expressive than RDF and RDFS, but like RDFS, it's also expressed in RDF. Using OWL, you can specify the cardinality of class membership and property values. For example, in OWL you could state that a company has only one CEO.

Another application of OWL is to logically merge data from different sources. For example, suppose that you've been using a property called `kb:organizationName` (specified using a shorthand namespace prefix of `kb:`) and that you also want to use another public RDF data store that uses a property called `europe:institutionName` for a list of European companies. You could write a converter to change properties, but you're better off leaving the two original data stores as-is and writing a statement that makes these properties equivalent in your application. You can do this easily using the reflexive OWL properties `owl:equivalentClass` and `owl:equivalentProperty`. Here is a single statement that would facilitate using data from two RDF sources with different ontologies:

```
kb:organizationName owl:equivalentProperty europe:institutionName .
```

Because `owl:equivalentProperty` is a reflexive property, you do not also need to state:

```
europe:institutionName owl:equivalentProperty kb:organizationName .
```

OWL features three variations, defined mostly by the type of logic supported for inferencing. You won't use most of OWL's extensions, but some knowledge of OWL will help you when I show you how to use the Protégé GUI application for working with RDF and RDFS data (see the section, "Working with the Protégé Ontology Editor"). Before you can use Protégé, you need to be able to convert N3 and N-Triple data to a format that Protégé can import. I cover this conversion in the next section.

Converting Between RDF Formats

Several web portals provide public conversion services that let you paste in text for RDF, RDFS, or OWL in one format and convert it to other formats. Here are two services that I use:

- `http://librdf.org/parse/`: Dave Beckett's conversion utility using his Redland RDF Library

- `http://sswap.info/format-converter.jsp`: The Simple Semantic Web Architecture and Protocol's conversion utility

You can also install Python tools from `http://www.w3.org/2000/10/swap/doc/cwm` to convert between RDF formats. Here is an example using the `cwm` command-line utility. Suppose you're given an N3 file like the `rdfs_sample_1.n3` example:

```
@prefix : <http://www.w3.org/2000/01/rdf-schema#> .
@prefix kb: <http://knowledgebooks.com/test#> .
@prefix rdf: <http://www.w3.org/1999/02/22-rdf-syntax-ns#> .

kb:Business a :Class;
     :label "Business";
     :subClassOf kb:Organization .

kb:Customer a :Class;
     :label "Customer";
     :subClassOf kb:Business .

kb:Organization a :Class;
     :label "Organization" .

kb:organizationName  a rdf:Property;
     :domain kb:Organization;
     :range <http://www.w3.org/2001/XMLSchema#stringstring> .
```

You can convert it to a legal RDF/RDFS/OWL XML file using this syntax:

```
cwm -n3 rdfs_sample_1.n3 -rdf > rdfs_sample_1.owl
```

This produces an XML format file that you will use for an example in the next section, when you use the Protégé GUI application to work with ontologies:

```
<rdf:RDF xmlns="http://www.w3.org/2000/01/rdf-schema#"
    xmlns:rdf="http://www.w3.org/1999/02/22-rdf-syntax-ns#">

    <Class rdf:about="http://knowledgebooks.com/test#Business">
        <label>Business</label>
        <subClassOf rdf:resource="http://knowledgebooks.com/test#Organization"/>
    </Class>

    <Class rdf:about="http://knowledgebooks.com/test#Customer">
        <label>Customer</label>
        <subClassOf rdf:resource="http://knowledgebooks.com/test#Business"/>
    </Class>

    <Class rdf:about="http://knowledgebooks.com/test#Organization">
        <label>Organization</label>
    </Class>

    <rdf:Property rdf:about="http://knowledgebooks.com/test#organizationName">
        <domain rdf:resource="http://knowledgebooks.com/test#Organization"/>
        <range rdf:resource="http://www.w3.org/2001/XMLSchema#stringstring"/>
    </rdf:Property>
</rdf:RDF>
```

I had promised that you would not be dealing directly with RDF XML files because they are more difficult to read. I am not breaking that promise: you need XML serialized RDF files to use Protégé, but you do not usually have to read them. Likewise, when you export XML-format data from Protégé, you can immediately convert it back to N3 or N-Triple format. But if you like to read and use the RDF/XML format, there is no reason not to use it.

Working with the Protégé Ontology Editor

For simple ontologies, it is fine to create a new N3-format file in a text editor and define namespaces, classes, and properties that you will need in your application. But a good alternative is to use a modeling tool such as Protégé to create the models you need. While it's not completely necessary for using the material in this book, I recommend that you take a few minutes right now to install Protégé from http://protege.stanford.edu and follow along with the examples in this section.

After you create your application's ontology models with either technique, you can use your ontology to write application programs that add data to an RDF data store (I'll address this in Chapter 5). Any ontologies you create will likely use models from existing namespaces such as Dublin Core (for modeling published works), Friend of a Friend (FOAF), and so on.

Figure 4-5 shows the Protégé UI after you load in the sample OWL file you created in the last section, when you used cwm to convert N3 to RDF/XML. Recent versions of Protégé are OWL-specific and expect file extensions to be OWL, so even if you are only modeling with RDF and RDFS (that is, not using OWL extensions), you still use OWL as the file extension when you want to import into Protégé.

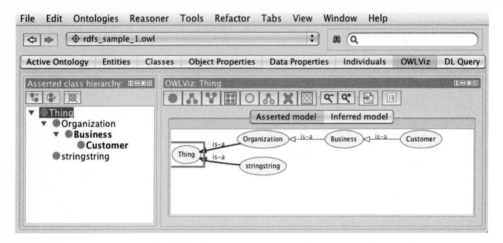

Figure 4-5. *The Protégé UI, which shows a class hierarchy that includes the three classes created earlier*

The file rdfs_sample_1.owl that you imported into Protégé contains three classes: Organization, Business, and Customer. Organization is a top-level class, so it shows as a subclass of Thing. Business is a subclass of Organization, and Customer is a subclass of Business. (In our simple model, customers are also businesses, not individual people.)

Later in this section you'll generate instances of classes in sample programs, and you'll also use both Protégé (in this chapter) and Graphviz (in Chapter 7) to visualize data. Protégé is a great tool for building ontologies, and I urge you to take the time to study one or more of the fine tutorials on the Protégé web site. I do want to show you one more thing in this section: using Protégé to create an instance of a class and then exporting the data as RDF/XML and converting it back to N3.

Figure 4-6 shows the Protégé UI right after I click the Individuals tab. Clicking the icon at the top-left corner of the Individuals pane on the left side of the screen creates a new individual of no specified class, and clicking the icon to the right of that deletes a selected instance.

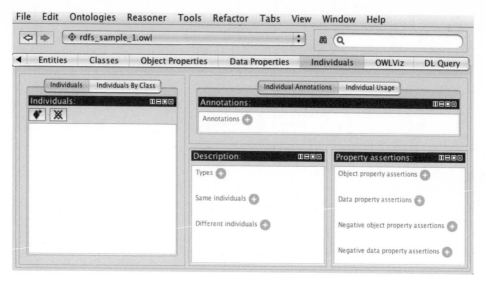

Figure 4-6. *Protégé showing the Individuals view*

Figure 4-7 shows the Protégé UI after you create an instance FreshVeggies and specify a type of Business for this instance by using the + button next to the Types label.

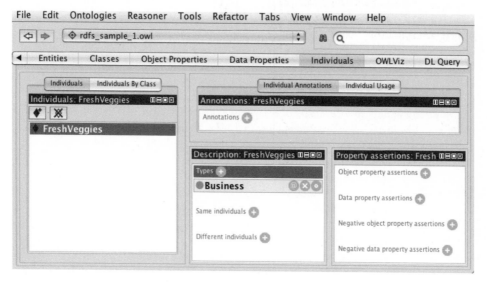

Figure 4-7. *Specifying that the instance FreshVeggies is of type Business*

After creating an instance of class Business, I saved the data as RDF/XML format into a file temp.rdf and then used the cwm command-line utility to create rdfs_sample_2.n3:

```
cwm -rdf temp.rdf -n3 > rdfs_sample_2.n3
```

If you look at the file `rdfs_sample_2.n3` in the directory `src/part2/data` (included in this book's source code, downloadable from the Apress web site), you will see the definition of a new namespace representing the file itself and a triple statement creating the individual:

```
@prefix rdfs_sample_1:
    <file:/Users/mark/ruby_scripting_book/src/part2/data/rdfs_sample_1.owl#> .

rdfs_sample_1:FreshVeggies a :Business .
```

When you are starting to use RDF and RDFS in your applications, you should try to combine the use of existing ontologies with those you create yourself. The use of Protégé or other modeling tools might not immediately be necessary, but I still encourage you to become familiar with at least one modeling tool. I expect that adoption of RDF as a data standard and the creation of public and private RDF data stores will be a slow but steady process. As more RDF data stores become available, more applications will take advantage of them. RDF data stores based on richer semantic models will be more valuable, which will produce a network effect: when we have more available data that conforms to rich semantic models (ontologies), then more intelligent software agents will be developed. I want you to be aware of tools such as Protégé even if you can get by without them in the short term.

Exploring Logic and Inference

I'll show you how to use the SPARQL RDF/RDFS/OWL query language in the next section. But processing SPARQL queries requires logic and inference, so I'll first provide an overview of logic. The study of logic, which is key to the study of philosophy, started in ancient Greece. The discipline involves the meaning of words and their relationships to the physical world and the world of ideas.

The study of logic includes the study of knowledge representation, which is the notion that human knowledge can be expressed via logic with mathematical notation. This branch of logic applies particularly to computer scientists and engineers, who take a more pragmatic approach to understanding how to encode and use knowledge in the software that they write. *Description logics* are knowledge-representation languages that strike a balance between being expressive and computationally efficient. Less expressive description logics are more efficient computationally, and vice versa. Different RDF data-storage systems utilize logic-based reasoning systems with different capabilities.

As you'll see in Chapter 5, the support for reasoning with RDF data conforming to RDFS and OWL models varies a great deal with different RDF storage systems. If you perform a web search for "descriptive logic reasoners," you will find about 20 different systems, most of which are produced by academic institutions but some of which are offered by private companies. You use logic reasoners to solve problems like these:

- *A father is also a parent.* (A statement that the `father` type is a subclass of the `parent` type.)

- *Ken is a father.* (A statement of a fact.)

- *Therefore, Ken is also a parent.* (A logical inference.)

Descriptive logic systems deal with relationships between classes (known as *terminology box* or *TBox* for historic reasons) and class instances (referred to as *assertional box* or *ABox*, also for historic reasons). Although not required, RDF data stores usually implement some form of *entailment*, which is the relationship between two sentences where the truth of one requires the truth of the other. You can see an example of entailment in the third bullet point, where a logical inference is used to show that Ken is a parent. In other words, entailment involves using RDF statements concerning things that are represented as true and deriving new statements that are also true.

Class membership in RDFS and OWL does not have to be explicit, defined by a statement such as instance A is in class B. Class B is likely defined in terms of the properties that class B instances have. If, through entailment, we can show that instance A has all of the properties for class B, then we say that A is *entailed* as belonging to class B. Here's a more concrete example: a class parent might be defined as instances with the property that they are either in the class mother or the class father. If we know that Ken is a father, then Ken also meets the requirements for being in class parent.

For the purposes of understanding and reusing the examples later in this book, I do not think that it is necessary for you to spend too much effort studying knowledge representation and description logics when you are just getting started using RDF and RDFS. I do want you, on the other hand, to have a road map for further study when you start using RDF data stores and SPARQL queries in your applications.

We are not done with logic and reasoning systems: in the next two sections, you will learn how to use SPARQL queries to access information in public RDF data stores. In Chapter 5, you will see some options for selecting RDF data stores and the types of reasoning systems that each choice provides. After getting some practical experience using RDF data in Chapter 5, you will revisit the topic of SPARQL queries and reasoning in Chapter 6.

Creating SPARQL Queries

The SPARQL query language is now a World Wide Web Consortium (W3C) standard, but there is much variation in the back-end reasoning systems that ultimately provide search results for SPARQL queries. I'll keep the discussion general in this section to provide you with enough knowledge to understand the programming examples in the next section. You will see sufficient SPARQL examples here and in the next three chapters, which will give you a good working knowledge of SPARQL by the end of Part 2.

■**Note** SPARQL is pronounced "sparkle."

SPARQL looks similar to SQL, but with an important difference: whereas you perform SQL queries against a relational database model, you perform SPARQL queries against graph data. The graph in Figure 4-4 shows the relationships between three classes, but in general RDF graphs also contain instance data: RDF graphs can represent relationships between classes, make statements about the relationships between data resources, and state simple facts. A SPARQL query matches parts of a graph represented by triples in an RDF data store and returns results matching variables in matching patterns. For brevity, SPARQL queries usually start by defining namespace-prefix abbreviations used later in the query. I'll spend the rest of this chapter on sample SPARQL queries, starting with one that uses the previous RDF examples:

```
PREFIX kb: <http://knowledgebooks.com/test#>
PREFIX rdf: <http://www.w3.org/1999/02/22-rdf-syntax-ns#>
SELECT ?organization
WHERE {
    ?organization rdf:type kb:Organization .
}
```

The preceding SPARQL query starts with namespace-prefix abbreviations and makes a select query to return matching values for the variable ?organization. Our previous example defined an instance of Business named FreshVeggies and specified Business as a subclass of Organization:

```
@prefix : <http://www.w3.org/2000/01/rdf-schema#> .
@prefix kb: <http://knowledgebooks.com/test#> .

kb:Business a :Class;
    :label "Business";
    :subClassOf kb:Organization .

kb:Organization a :Class;
    :label "Organization" .
kb:FreshVeggies  a :Business .
```

You would expect to get a query result matching the variable ?organization equal to FreshVeggies. (You'll see in Chapter 5 that this assumption is not always correct because some data stores do not support RDFS type inference.) This example is simple, with only one matching variable and only one matching statement in the WHERE clause. As you will see in more examples using real RDF data stores, you will usually use more than one matching statement.

I will introduce SPARQL features as I need them in the rest of Part 2, including:

- Using regular expressions to match on string literal values (almost always the object part of a triple)

- Filtering on numeric values or ranges on numeric literals (also almost always the object part of a triple)

- Using blank nodes in query results (almost always to represent a specific subject in multiple matched triples)

Even with examples of this additional functionality, I'll be using a small subset of SPARQL in this book. I determined which subset to discuss by implementing the programming examples and then introducing what you need to understand and reuse the examples.

Accessing SPARQL Endpoint Services

For the examples in this section, I'll use a Ruby library written by Dan Brickley and patterned after a Python client by Ivan Herman. You can find Dan's original code on the Web (http://svn.foaf-project.org/foaftown/2008/ruby-sparql-client/). I simplified his code and made a few changes; you'll find the simplified version in the file src/part2/pure_ruby_sparql_client/sparql_client.rb (downloadable from the Apress web site). The Ruby SPARQL client library processes a SPARQL query (a string value), calls a remote SPARQL endpoint service, and then returns an XML payload that you need to process in your application. A SPARQL endpoint is a web-service wrapper for an RDF data store; the endpoint accepts SPARQL queries and returns query results based on the RDF in the local data store. I usually prefer to use N-Triple and N3 formats rather than RDF/XML, but using an XML parser as I did in Chapter 1, I can easily extract the necessary information.

You'll use two public SPARQL endpoints in our examples using this library. (In the next two chapters, you'll create and use our own SPARQL endpoints.) I use the same process for each of the two public endpoint examples. I start by performing a "match anything" query to see what information is available. This query specifies three match variables (?s, ?p, and ?o) with a "match anything" WHERE clause, but it limits the number of results to 100:

```
SELECT ?s ?p ?o WHERE {
  ?s ?p ?o .
}
LIMIT 100
```

Now I'll show you an example of a more restrictive query. Suppose you want to find all triples with a predicate of rdf:type. This query would return subjects and their types in an RDF data store:

```
SELECT ?s ?o WHERE {
  ?s rdf:type ?o .
}
LIMIT 100
```

Using a public SPARQL endpoint starts with exploring the available data, determining what predicates are used and which might be useful for a specific application, and developing one or more custom SPARQL queries to fetch useful data.

The following listing shows the implementation of the pure Ruby client library:

```
# author: Dan Brickley
# patterned after a Python client by Ivan Herman
# modifications: Mark Watson
# license: W3C SOFTWARE NOTICE AND LICENSE

require 'open-uri';  require 'rexml/document';  require 'cgi'
```

```ruby
module SPARQL
  class SPARQLWrapper
    attr_accessor :baseURI, :_querytext, :queryString, :URI, :retval
    def initialize(baseURI)
      self.baseURI = baseURI;  self._querytext  = []
      self.queryString = "SELECT * WHERE{ ?s ?p ?o }"
      self.retval = nil; self.URI    = ""
    end
    def fetch_uri # Return the URI to be sent to the SPARQL endpoint.
      self._querytext.push(["query",self.queryString])
      begin
        esc = ""
        self._querytext.each {
          |a| esc += a[0] + "=" + CGI.escape(a[1]) +"&"
        }
        self.URI= self.baseURI + "?" + esc
      rescue
        puts "Error: url escaping... #{$!} self.URI: #{self.URI}\n"
      end
      return self.URI
    end
    def query()  # Execute the query.
      QueryResult.new(open(self.fetch_uri).read) rescue puts "Error query. #{$!} "
    end
  end

  class QueryResult
    attr_accessor :response
    # @param response: HTTP response stemming from web service call
    def initialize(response)
      self.response = response
    end
    # Return the URI leading to this result
    def geturl
      return self.response.geturl
    end
    # Return the meta-information of the HTTP result
    def info
      return self.response.info
    end
    # Convert an XML result into a dom tree.
    def convertXML
      return REXML::Document.new(self.response)
    end
  end
end
```

This library wraps an endpoint URL and a SPARQL query, then Common Gateway Interface (CGI) encodes it and makes a service call to the endpoint service. The response is a string representation of the SPARQL XML response. Here you used the REXML library to create a DOM object from this string representation.

In the next two sections, I'll show you two examples of using free endpoint services on the Web.

> **Caution** Many free SPARQL endpoints for public RDF repositories are part of research projects, so they might not always be up and running. In most cases, I find that interrupted services are soon back online. You'll see in Chapter 7 how to create your own SPARQL endpoint service as a pure Ruby application. And in Chapter 5, you'll use RDF repositories implemented in Common Lisp (AllegroGraph) and Java (Sesame) that you can download from the Web and install locally.

Using the Linked Movie Database SPARQL Endpoint

A web search on "RDF SPARQL public endpoints" quickly yields many public information sources. In this section, you'll use the Linked Movie Database maintained by the University of Toronto as a research project. The SPARQL endpoint URI for this service is `http://data.linkedmdb.org/sparql`, and you can start with a SPARQL query like this one to see some available data:

```
SELECT ?p  WHERE {
  ?s ?p ?o .
}
LIMIT 30

SELECT distinct ?p  WHERE {
  ?s ?p ?o .
}
LIMIT 30
```

The first query matches any triples in the RDF data store; it can help you quickly get a rough idea of what information is available. The second query uses the `distinct` keyword to remove duplicate query results and returns a maximum of 30 predicates used in the triple store. From the first query, you can see that some triples contain objects that are string literals for movie titles. You can fetch up to ten titles along with a performance-resource URI and a performance-type predicate using this new query (assuming that the variable `search_phrase` contains words to search for):

```
SELECT ?s ?p ?o WHERE {
  ?s ?p ?o FILTER regex(?o, \"#{search_phrase}\")
}
LIMIT 10
```

You get XML result elements like this:

```
<result>
   <binding name='s'>
     <uri>http://data.linkedmdb.org/resource/performance/581 </uri>
   </binding>
   <binding name='p'>
    <uri>http://data.linkedmdb.org/resource/movie/performance_film
    </uri>
   </binding>
   <binding name='o'>
     <literal>A Fistful of Dollars</literal>
   </binding>
</result>
```

Using the client library, you can print out the results with a short script:

```
require "sparql_client"

ENDPOINT="http://data.linkedmdb.org/sparql"

def search_movies search_phrase
  qs="SELECT ?s ?p ?o WHERE {
    ?s ?p ?o FILTER regex(?o, \"#{search_phrase}\")
  }
  LIMIT 10"
  sparql = SPARQL::SPARQLWrapper.new(ENDPOINT)
  sparql.setQuery(qs)
  ret = sparql.query
  xmldoc = ret.convertXML
  xmldoc.each_element('//result') {|result|
    result.each_element("//binding") {|binding|
      binding.each_element('uri') {|uri|
        puts "#{binding.attribute('name')} : #{uri.text}"
      }
      binding.each_element('literal') {|literal|
        puts "#{binding.attribute('name')} : literal: #{literal.text}"
      }
    }
  }
end
```

Here is some sample output showing the subject, predicate, and object for one triple:

```
s : http://data.linkedmdb.org/resource/performance/581
p : http://data.linkedmdb.org/resource/movie/performance_film
o : literal: A Fistful of Dollars
```

If you follow the URI for the subject, you will see that it refers to actor Clint Eastwood. While it would be possible to add Ruby code to go back and make additional queries to fetch the names of all the actors in this movie, you're better off letting the RDF data service do the heavy lifting and using a compound query (slightly analogous to performing a join in a SQL query). The following query returns movie titles and actors for movie titles matching the search keywords:

```
SELECT ?title ?actor WHERE {
    ?s ?p ?title FILTER regex(?title, \"#{search_phrase}\") .
    ?s <http://data.linkedmdb.org/resource/movie/performance_actor> ?actor .
  }
  LIMIT 20
```

You get results like this:

```
<result>
    <binding name='title'>
      <literal>A Fistful of Dollars</literal>
    </binding>
    <binding name='actor'>
      <literal>Clint Eastwood</literal>
    </binding>
</result>
```

You can substitute the XML-parsing code from the last example with this:

```
xmldoc.each_element('//result') {|result|
  title = actor = nil
  result.each_element("binding") {|binding|
    binding_name = binding.attribute('name').to_s.strip
    binding.each_element('literal') {|literal|
      actor = literal.text if binding_name == 'actor'
      title = literal.text if binding_name == 'title'
    }
  }
  puts "#{title}\t:\t#{actor}" if title && actor
}
```

You now see that the following actors worked in the movie *A Fistful of Dollars*:

```
A Fistful of Dollars    :    Clint Eastwood
A Fistful of Dollars    :    Marianne Koch
A Fistful of Dollars    :    Gian Maria Volonta
A Fistful of Dollars    :    Antonio Prieto
A Fistful of Dollars    :    Mario Brega
A Fistful of Dollars    :    Wolfgang Lukschy
A Fistful of Dollars    :    Seighardt Rupp
```

```
A Fistful of Dollars    :    Josef Egger
A Fistful of Dollars    :    Benito Stefanelli
A Fistful of Dollars    :    Jose Calvo
A Fistful of Dollars    :    Peter Fernandez
```

Most public SPARQL endpoints on the Web also provide a web-browser interface that helps you quickly determine which property names and the like that you need for building queries. I dream of writing software systems that explore RDF data graphs automatically, follow useful references, and summarize the results. You can build up near-term practical applications with custom programming by treating multiple RDF stores as a distributed database, manually exploring available schemas and data, and building up SPARQL graph queries to find useful information.

Using the World Fact Book SPARQL Endpoint

The Free University of Berlin supports several public SPARQL endpoints for research purposes. Whenever you experiment with a free RDF repository, remember that someone is providing a free service to you. You can minimize their server costs by using the SPARQL LIMIT value to limit the number of matches returned for a query. You'll now use the Free University of Berlin's SPARQL endpoint for CIA World Fact Book data: http://www4.wiwiss.fu-berlin.de/factbook/sparql.

You can use the following query to get 30 unique predicates in the repository by editing the example file src/part2/pure_ruby_sparql_client/test_query_world_factbook.rb and changing the query (variable qs) to:

```
SELECT distinct ?p  WHERE {
  ?s ?p ?o .
}
LIMIT 30
```

When I have my SPARQL queries and client code working, I remove the LIMIT value to get all possible answers. A typical result element looks like this (I shortened the URI to fit in one line):

```
<result>
  <binding name="p">
    <uri>http://wiwiss.fu-berlin.de/factbook/ns#lifeexpectancyatbirth_female</uri>
  </binding>
</result>
```

Here is a Ruby script to print all predicates in the repository:

```
require "sparql_client"

qs="SELECT distinct ?p  WHERE {
  ?s ?p ?o .
```

```
}
LIMIT 10"

endpoint="http://www4.wiwiss.fu-berlin.de/factbook/sparql"

sparql = SPARQL::SPARQLWrapper.new(endpoint)
sparql.setQuery(qs)
begin
   ret = sparql.query
   xmldoc = ret.convertXML
   xmldoc.each_element('//result/binding/uri') {|uri|
     puts uri.text
     }
end
```

I chose the following three predicates to use in the second example of a compound SPARQL query:

```
http://www4.wiwiss.fu-berlin.de/factbook/ns#countryname_conventionalshortform
http://www4.wiwiss.fu-berlin.de/factbook/ns#capital_name
http://www4.wiwiss.fu-berlin.de/factbook/ns#population_total
```

Here is the SPARQL query with the triples split over three lines each (I shortened the URIs to fit on one line):

```
SELECT ?name ?capital ?population  WHERE {
  ?country
     <http://wiwiss.fu-berlin.de/factbook/ns#countryname_conventionalshortform>
     ?name .
  ?country
     <http://wiwiss.fu-berlin.de/factbook/ns#capital_name>
     ?capital .
  ?country
     <http://wiwiss.fu-berlin.de/factbook/ns#population_total>
     ?population .
}
LIMIT 30
```

So far, you have been specifying complete URIs in SPARQL SELECT statements. Here is the same query using a namespace-prefix abbreviation:

```
prefix wf: <http://www4.wiwiss.fu-berlin.de/factbook/ns#>
select ?name ?capital ?population  where {
  ?country
     wf:countryname_conventionalshortform
     ?name .
  ?country
```

```
      wf:capital_name
      ?capital .
  ?country
      wf:population_total
      ?population .
}
limit 30
```

You might find that some SPARQL endpoints for public RDF repositories do not support namespace prefixes, so it is important to know how to use either query form. Here is some sample output:

```
Bermuda    : Hamilton    : 66163
Belgium    : Brussels    : 10392226
```

The Free University of Berlin also supports a public SPARQL endpoint for DBpedia, which is a Semantic Web version of Wikipedia articles.

Wrapup

I'll continue to demonstrate Semantic Web technologies in the remainder of this book. My goals in this chapter were to make you feel comfortable with RDF graph-based data and to give you some tools to get started on your own applications. Although it is beyond the scope of this book, I would like to share some of my research in following references in RDF graphs. Consider the graph in Figure 4-8. This figure represents a simplified RDF graph as an example of how a software agent might start at a known URI for a repository-directory system and process queries such as "where can I buy carrots?" In order to solve a problem like this, you would need knowledge-representation statements such as "if a business sells X, then you can buy X from the business." Then a software agent could start searching for all triples with a predicate representing "sell" and for objects containing a literal string value for "carrot" or for URIs representing the concept of a carrot. While it is easy for a human to browse the Web and follow links, there are many challenges inherent in trying to implement this behavior in software agents. Definitely an open research problem!

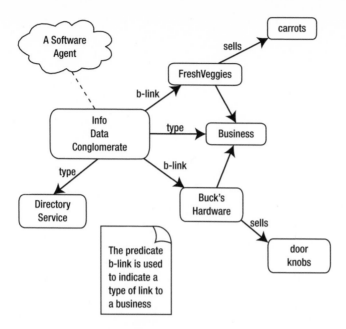

Figure 4-8. *A simplified RDF graph showing a business directory*

Some RDF data sources on the web have custom (non-SPARQL) interfaces. As I write this chapter, the Freebase.com semantic-data web portal provides a REST web-service interface for fetching RDF representations of its articles. You'll use Freebase in Chapter 11. You'll also use the Semantic Web version of Wikipedia (DBpedia) in Chapter 11.

You should now have a basic understanding of RDF, RDFS, and SPARQL. The introduction in this chapter concerning the use of SPARQL clients will be continued in Chapters 6 and 7. In the next chapter, I'll show you several options for implementing and using RDF data stores.

■ ■ ■

## CHAPTER 5

Delving Into RDF Data Stores

Efficiently storing and querying RDF data is key to the Semantic Web and Web 3.0. In this chapter, I will show you some of the practical issues of using RDF by working through the setup and use of three RDF data stores. In Chapter 6, I will cover the technical details of using RDF with SPARQL queries. Chapter 7 contains examples for publishing RDF data, and in Chapter 10 you'll see how you can use RDF to store relations between data extracted from different web sites.

For each data store, you'll study examples of using Ruby clients. I will cover Redland, Sesame, and Franz's AllegroGraph systems. Redland is an open source tool that uses either the Apache License, Version 2.0 or the GNU Lesser General Public License (LGPL), based on your choice. Sesame is an open source tool that's released under a BSD-style license. AllegroGraph is a commercial product; a limited-use version (limited to 50 million RDF triples) is available under license for free use. There are other good options, but I determined that these three good choices are more than adequate. Other options include commercial database products from IBM and Oracle that support RDF and OpenLink Virtuoso (either the open source or commercial version).

You saw in Chapter 4 that RDF data stores can form a web-wide system of distributed databases supporting public data. Your goal in this chapter is to learn how to participate in this system by providing your own data publicly to the Web at large or privately inside your company's local network. There are two skill sets for you to learn: publishing your own semantic data and consuming semantic data that other people publish.

■**Tip** You can either install the software packages used in this chapter on your own development system or use the custom Amazon EC2 AMI instance that I created for this book (see Appendix A for details). You can save a lot of effort using the EC2 AMI instance.

Installing Redland RDF Ruby Bindings

The Redland RDF Library, written and supported by Dave Beckett, is available at http://librdf.org. This library is implemented in C with bindings for Python, Ruby, Perl, and PHP. Redland supports several back-end data stores: Berkeley DB, SQLite 3, MySQL, PostgreSQL, and in-memory. On Linux and Mac OS X systems, Redland will build using all available back ends that it finds. But using the Berkeley DB system, which Oracle purchased from Sleepycat Software, is the best back-end choice for large RDF data stores. Berkeley DB is released under the GPL and is available in commercial versions if the GPL does not work for your projects.

You can download the Redland RDF Library from `http://librdf.org` and you can find specific installation details for the Ruby Redland language bindings from `http://librdf.org/docs/ruby.html`. In addition to the Redland RDF Library, you need to download the Raptor RDF Parser Library and the Rasqal RDF Query Library. Probably the easiest install for experimenting with Redland is to configure it specifically for SQLite 3 (assuming that you already have SQLite 3 installed). After downloading the compressed TAR files for Raptor, Rasqal, and Redland, follow these steps to install them to the `/usr/local/` directory:

```
cd raptor-1.4.18
./configure
make
sudo make install
cd ../rasqal-0.9.16
./configure
make
sudo make install
cd ../redland-1.0.8/
./configure --with-ruby --with-sqlite=3
make
sudo make install
cd ../redland-bindings-1.0.8.1/
./configure --with-ruby
 make
cd ruby
sudo make
```

Note These instructions work for me on Mac OS X and Linux. If you use Windows, you can find some old versions of Redland prebuilt at `http://download.librdf.org/binaries/win32/`. That said, if you must use Windows, then you might want to skip ahead to the next section and use Sesame.

On my system, I like to install Ruby from source in my home directory, so I specify the location of my custom Ruby install by using the `--with-ruby` option:

```
./configure --with-ruby=/Users/markw/bin/ruby --with-sqlite=3
./configure --with-ruby=/Users/markw/bin/ruby
```

Warning As I am writing this (February 2009), Redland does not build a Ruby binding for Ruby 1.9.x. I used Ruby 1.8.6 for the examples in this section.

When you run the first `configure` command, you will see a summary of the available storage back ends and the types of triple stores that you can use:

```
Redland build summary:

  Berkeley/Sleepycat DB   :
                  Version 4.6 (library db-4.6 in /usr/local/BerkeleyDB.4.6/lib)
  Triple stores available :
                  file hashes(memory) trees hashes(bdb 4.6) mysql(5.1.22-rc)
                                    sqlite(3.5) postgresql(8.3.4)
  Triple stores enabled   : memory file hashes trees mysql sqlite postgresql
```

After you finish the install, you will be working in the `redland-bindings-1.0.8.1/ruby` directory, and you can test the install with this:

```
ruby example.rb file:test/ical.rdf rdfxml
```

Redland is an efficient RDF data store (especially when you use Berkeley DB for back-end storage) and is a good choice for embedded Semantic Web applications. While Redland does support SPARQL queries, it does not support RDFS inferencing; for example, `rdfs:subClassOf` and `rdfs:subPropertyOf` are not supported. Because of this restriction, I recommend you use Redland for simpler RDF applications that require a fast RDF store with SPARQL-query capability but that don't require RDFS inferencing. For applications that require RDFS or OWL inferencing (reasoning), I use Sesame and AllegroGraph. In Chapter 6, I'll discuss inferencing and reasoning with Sesame in some depth, but the discussion will also apply if you use AllegroGraph.

The file `src/part2/redland_demo.rb` contains the complete code for a sample use of Redland; I'll now use this code for the following examples. (You can find the code samples for this chapter in the Source Code/Download area of the Apress web site at http://www.apress.com.) This Ruby client example differs from what you'll see when you use Sesame and AllegroGraph web services later, because the Redland Ruby client uses an embedded RDF data store and the Sesame/AllegroGraph example clients use REST-style web-service APIs.

The following code snippets are all from the file `redland_demo.rb` that I derived from the `example.rb` test file in the `redland-bindings` distribution:

```
require 'rdf/redland'

PARSER_RDF_TYPE="ntriples"

STORAGE=Redland::TripleStore.new("hashes", "test",
                                 "new='yes',hash-type='bdb',dir='data'")
raise "Failed to create RDF storage" if !STORAGE
```

The possible values for `PARSER_RDF_TYPE` are `ntriples`, `n3`, and `rdfxml`. The first argument in `Redland::TripleStore.new` can be `memory`, `file`, `hashes`, `mysql`, `sqlite`, or `postgresql`. The second argument is used as a root file name for disk-based, back-end data stores. The third argument is a string encoding of options. The default value for the `new` option is `no`, which indicates that you're using a previously created data store (in this example code snippet, the value is yes because I am creating a new data store). The hash-type value of bdb refers to a Berkeley DB back end, and a value of memory would specify a high-performance, in-memory hash library. The dir option specifies the directory path where persistent disk files are stored.

Note I have used only in-memory and Berkeley DB back ends.

The following methods are defined in the file redland_demo.rb:

```
def load_rdf_data uri_string; ...; end
def query sparql_query; ...; end
```

In these helper functions, I packaged the Redland-bindings example code for reading RDF stores from any URI and for performing queries. To see the implementation, you can look at the source in the src/part2/redland_demo.rb file.

I'll use sample_news.nt, a file in RDF N-Triple format, in the following examples. (You can check out src/part2/data to look at this sample file.) Here are the first two triples defined in this file:

```
<http://kbsportal.com/oak_creek_flooding />
    <http::://knowledgebooks.com/ontology/#storyType>
    <http::://knowledgebooks.com/ontology/#disaster> .

<http://kbsportal.com/oak_creek_flooding />
    <http::://knowledgebooks.com/ontology/#summary>
    "Oak Creek flooded last week affecting 5 businesses" .
```

I use two of my own domain names, knowledgebooks.com and kbsportal.com, in these examples. The following code snippet loads this example RDF data file and performs two SPARQL queries:

```
load_rdf_data("file:data/sample_news.nt")
query(" SELECT ?s ?o
            WHERE { ?s <http::://knowledgebooks.com/ontology/#storyType> ?o } ")
query(" SELECT ?s ?o
            WHERE { ?s <http::://knowledgebooks.com/ontology/#summary> ?o } ")
```

Output from the first query looks like this:

```
<http://kbsportal.com/trout_season />
<http::://knowledgebooks.com/ontology/#recreation>

<http://kbsportal.com/trout_season />
<http::://knowledgebooks.com/ontology/#sports>
```

Output from the second query looks like this:

```
<http://kbsportal.com/bear_mountain_fire />
"The fire on Bear Mountain was caused by lightning"

<http://kbsportal.com/oak_creek_flooding />
 "Oak Creek flooded last week affecting 5 businesses"
```

I will use Redland in Chapter 7 to develop a pure Ruby (except for the Redland RDF Library itself) RDF web portal that supports SPARQL clients. The web portal we'll develop in Chapter 7 is a lighter-weight alternative to the Sesame libraries and RDF web portal that we will use in the next section.

Using the Sesame RDF Data Store

I have been using the Sesame library for more than five years, embedding it in Java applications and running it as a web service in an Apache Tomcat web container. Because this book covers mostly Ruby development, I will primarily use Sesame in a server configuration with Ruby clients. In the next section, I will also show you how to use Sesame embedded directly in JRuby scripts and applications. If you are uncomfortable installing Java software, you can quickly read through this section and use another alternative RDF data store such as AllegroGraph.

If you don't yet have Java installed, download a JDK installer from `http://java.sun.com/`. If you do have Java installed, you need to download and install both a Tomcat web container and a Sesame SDK from the following locations:

- *Tomcat version 6.x*: `http://tomcat.apache.org/`

- *Sesame version 2.x*: `http://www.openrdf.org`

Although Tomcat is open source, you probably want to download only the binary core distribution initially, and not the source code. A quick way to install Tomcat with Sesame is to download a binary Tomcat distribution, download the Sesame SDK, unzip both, and then copy the two files from the Sesame distribution `war` directory (`openrdf-sesame.war` and `openrdf-workbench.war`) to the Tomcat `webapps` directory. You can follow these steps:

1. `cp openrdf-sesame-2.2.4/war/openrdf-sesame.war apache-tomcat-6.0.18/webapps/`

2. `cp openrdf-sesame-2.2.4/war/openrdf-workbench.war apache-tomcat-6.0.18/webapps/`

3. `cd apache-tomcat-6.0.18/bin`

To start the service running interactively (so you immediately see any error messages) from the Tomcat `bin` directory, run the following command:

```
./catalina.sh run
```

■**Tip** If you are running on Windows, use this command instead: `catalina.bat run`.

You can access the service using two URLs (assuming that you are running Tomcat on your local computer):

- `http://localhost:8080/openrdf-sesame`: This is the Administration web application. Click the System tab to see where Sesame is storing the RDF repository on your local fie system; note this location because you might want to delete this data when you are done.

- `http://localhost:8080/openrdf-workbench`: This is the Workbench application for using repositories. Click the New Repository link and create a repository of type **Native Java Store RDF Schema and Direct Type Hierarchy**. Set the ID to **101** and the name to **business**.

Figure 5-1 shows part of the Sesame Administration web interface.

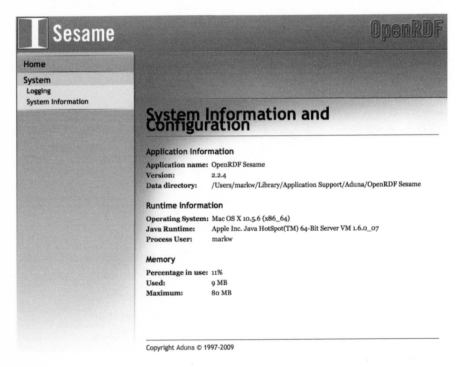

Figure 5-1. *Sesame administration web application*

Using the Workbench web interface (see Figure 5-2), click the Modify ➤ Add link and load the N3 data file from the directory src/part2/data/rdfs_business.n3. To see the data that you just loaded, click the Explore ➤ Types link to see a list of all types in the repository. You can start by clicking the link <http://knowledgebooks.com/test#Business> on the Workbench web application's Types page (see Figure 5-3) to see the triples that contain this URI for our test class Business. The Workbench web interface is very good; spend some time exploring the data that you just loaded from my sample N3 file.

I seldom use the Sesame Administration web application (see Figure 5-1) unless something goes wrong. Clicking the Logging link on the left side of the screen shows information about web-service requests and internal processing by the Sesame libraries. The System Information link beneath the Logging link shows detailed information about the Java runtime environment.

The Sesame Workbench web application is shown in Figures 5-2, 5-3, and 5-4. I use this web application to bulk-load data into repositories, manage repositories, and interactively explore and use SPARQL queries on selected repositories.

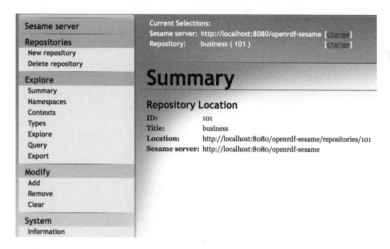

Figure 5-2. *Sesame Workbench web application showing a repository summary after a repository link is clicked*

Sesame is close to ideal for publishing RDF data. The web interface makes it easy to manually publish RDF data to a remote server, and you'll see how to automatically publish data using client programs.

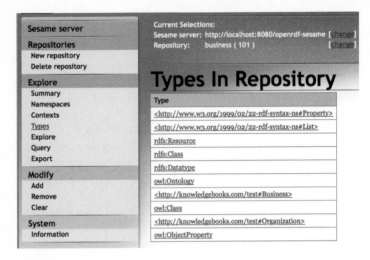

Figure 5-3. *The Types page of the Sesame Workbench web application*

The Types In Repository view shown in Figure 5-3 is a useful starting point for viewing triples that contain a specific RDF type. I clicked the link <http://knowledgebooks.com/test#Business> in Figure 5-3 to get the display shown in Figure 5-4.

Super Classes
- rdfs:Resource
- <http://knowledgebooks.com/test#Organization>

Sub Classes
- <http://knowledgebooks.com/test#Customer>

Subject	Predicate	Object
<http://knowledgebooks.com/test#Business>	<http://www.w3.org/1999/02/22-rdf-syntax-ns#type>	rdfs:Resource
<http://knowledgebooks.com/test#Business>	<http://www.w3.org/1999/02/22-rdf-syntax-ns#type>	rdfs:Class
<http://knowledgebooks.com/test#Business>	<http://www.w3.org/1999/02/22-rdf-syntax-ns#type>	owl:Class

Figure 5-4. *The triples for the test class Business shown in the Sesame Workbench*

Interactively running the Sesame web interfaces was a good start. However, the real utility in using Sesame in your distributed applications is using the server as a black box and using client applications (Ruby apps, for our examples) to load data into the repository and perform queries. If you are writing Java clients, you would probably choose to use the supplied Java client libraries. In Ruby, you'll use the HTTP communication protocol for Sesame 2.x that is also supported by AllegroGraph. The Sesame protocol uses the REST architectural style. You'll use the following URI patterns (copied from the online Sesame system documentation):

```
<SESAME_URL>
    /protocol          : protocol version (GET)
    /repositories      : overview of available repositories (GET)
        /<REP_ID>      : query evaluation on a repository (GET/POST)
            /statements : repository statements (GET/POST/PUT/DELETE)
```

```
/contexts     : context overview (GET)
/size         : #statements in repository (GET)
/namespaces   : overview of namespace definitions (GET/DELETE)
    /<PREFIX> : namespace-prefix definition (GET/PUT/DELETE)
```

You can find all the following Ruby client code snippets in the source file src/part2/ sesame_rest_test_1.rb (downloadable from the Apress web site). You will need to install the JavaScript Object Notation (JSON) gem and restclient gem using gem install json and gem install restclient, respectively.

■**Tip** The JSON gem requires native extensions to be built and installed. If this does not work on your system, use the slower "pure Ruby" version: gem install json_pure.

Start this example of making a REST web-service call to Sesame by requiring a few gems and fetching the repository-protocol version:

```
require 'rubygems'
 require 'restclient'
require 'cgi'
 require 'json'
require 'pp'

protocol = RestClient.get("http://localhost:8080/openrdf-sesame/protocol")
puts protocol
```

The preceding code snippet prints out "4" as the protocol value. The value "4" specifies the version of the Sesame REST APIs. You could get a list of repositories using this code:

```
repositories =
  RestClient.get("http://localhost:8080/openrdf-sesame/repositories",
                    :accept => 'application/sparql-results+json')
pp repositories
repo_json = JSON.parse(repositories)
```

The repository that you just created using the Workbench web interface has an ID of 101, so you can use this directory for submitting a SPARQL query and getting the results in JSON format:

```
query ="SELECT ?subject ?predicate WHERE {
    ?subject ?predicate <http://knowledgebooks.com/test#Business> .
}"

uri = "http://localhost:8080/openrdf-sesame/repositories/101?query="
uri += CGI.escape(query)
payload = RestClient.get(uri, :accept => 'application/sparql-results+json')
```

```
json_data = JSON.parse(payload)
pp json_data
```

Here is a very small part of the SPARQL query results:

```
{"results"=>
  {"bindings"=>
    [{"predicate"=>
        {"type"=>"uri",
         "value"=>"http://www.w3.org/1999/02/22-rdf-syntax-ns#type"},
      "subject"=>
        {"type"=>"uri",
         "value"=>"http://knowledgebooks.com/test#HanksHardware"}},
     {"predicate"=>
        {"type"=>"uri",
          "value"=>"http://www.openrdf.org/schema/sesame#directSubClassOf"},
      "subject"=>
        {"type"=>"uri", "value"=>"http://knowledgebooks.com/test#Customer"}}]},
 "head"=>{"vars"=>["subject", "predicate"]}}
```

These results state that HanksHardware is a type of Business and that Customer is a subclass of Business.

The hash table returned by JSON.parse has two keys: results and head. The head contains another hash table with values equal to the matching variable names in the SPARQL query.

The restclient Ruby gem has a nice feature: you can set HTTP request headers by adding a second optional argument to the class method RestClient.get. You requested that the Sesame web services return JSON data, but you have three format options:

```
:accept => 'application/sparql-results+json'
:accept => 'application/sparql-results+xml'
:accept => 'application/x-binary-rdf-results-table'
```

The last option is a binary format that might be appropriate for higher performance if you expect very large payloads. For this example, you'll only use the JSON and XML text formats.

In addition to using the Sesame Workbench web interface for adding large data sets to a repository, you can also use HTTP POST to add data:

```
s = "<http://knowledgebooks.com/test#WorkType>
     <http://www.w3.org/1999/02/22-rdf-syntax-ns#type>
     <http://www.w3.org/2002/07/owl#Class> ."
uri2 = "http://localhost:8080/openrdf-sesame/repositories/101/statements"
sRestClient.post(uri2, s, :content_type => 'application/x-turtle')
```

REST web-service calls to the repository use the HTTP actions GET, POST, PUT, and DELETE. You use GET to fetch triple data using SPARQL queries, you use PUT to update data that is already in the repository, and you use POST to add new data to a repository. You use DELETE to remove data matching a SPARQL query from the repository. The Sesame REST APIs are fully documented in Chapter 8 of the online Sesame system documentation.

The combination of the Sesame libraries and the Sesame web applications is a powerful tool for building distributed RDF data stores and remote SPARQL client-based Web 3.0 applications. In Chapter 7, I will develop an alternative using JRuby: one of the example programs will be a lightweight web-portal wrapper around the core Sesame libraries. You'll learn to use Sesame in embedded applications in the next section.

Note If you have not installed JRuby, follow the directions at `http://wiki.jruby.org/wiki/ Getting_Started`.

Embedding Sesame in JRuby Applications

You saw how to use Redland as an embedded RDF data store, much as you use SQLite as an embedded relational database. While I almost always use Sesame as a networked service, as in the last section, you can also embed it in Java and JRuby programs. I will show you how to wrap the Sesame Java APIs for use in JRuby scripts.

There are two different approaches to reusing Java libraries in JRuby scripts. One method directly loads JAR files, include classes, and so on. The other technique is to write a wrapper in Java with specific APIs that you want to use in JRuby. Writing a Java wrapper class takes more work, but the advantage is that you can create simpler APIs for accessing complex Java class libraries. While I will use the second approach of writing a wrapper class, first we'll explore an example for the first approach: directly using Java classes in JRuby.

To directly use Java classes in JRuby, you start by loading a JRuby module that loads Java standard classes, and then you load every JAR file in a subdirectory lib. Start by changing the directory to the Sesame SDK directory, which you should have on your system if you have already set up Tomcat and Sesame web services:

```
cd openrdf-sesame-2.2.4
```

You can then run the JRuby interactive shell `jirb` and input the following statements:

```
include Java
Dir["lib/*.jar"].each { |jar| require jar }
```

Next, you need to include Java classes in nonstandard libraries that you just loaded as JAR files:

```
include_class org.openrdf.sail.memory.MemoryStore
include_class org.openrdf.sail.inferencer.fc.ForwardChainingRDFSInferencer
include_class org.openrdf.repository.sail.SailRepository
```

You can now directly use these classes in Ruby code:

```
ms = MemoryStore.new
fci = ForwardChainingRDFSInferencer.new(ms)
my_repository = SailRepository.new(fci)
```

For more details on this first approach of directly using the Sesame libraries, you can read the Sesame documentation. However, for the rest of this section I will take the second approach: writing a Java wrapper class. You'll want to check out the src/part2/jruby_sesame/ rsesame.jar file for the purposes of this example (this chapter's source code is downloadable from the Apress web site). Figure 5-5 shows the Unified Modeling Language (UML) class diagram for the wrapper.

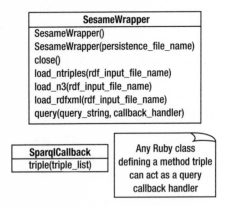

Figure 5-5. *UML class diagram for the Sesame wrapper class and callback interface*

Using the wrapper class, you can load data files into an RDF repository and make queries—as you did when using the Sesame web services—except that Sesame is now embedded in your Ruby scripts and applications. The file src/part2/jruby_sesame/test.rb shows how to use the wrapper class:

```ruby
include Java
require "rsesame.jar"
include_class "SesameWrapper"

sw = SesameWrapper.new
sw.load_n3("../data/rdfs_business.n3")

# Use duck typing: define a class implementing the method "triple" :
class TestCallback
  def triple result_list # called for each SPARQL query result
    puts "Matching subject: #{result_list[0]}\n        predicate: #{result_list[1]}"
  end
end

callback = TestCallback.new

sparql_query =
"PREFIX kb:  <http://knowledgebooks.com/test#>
SELECT  ?subject ?predicate ?object2
WHERE {
```

```
    ?subject ?predicate kb:Amazon .
}"
```

```
sw.query(sparql_query, callback)
```

In this example, I am making a SPARQL query that returns matching values for subjects and predicates, so my callback handler assumes that the triple list contains two elements. If you run this example, you will notice a lot of diagnostic output from the Sesame libraries; the following output from the previous snippet does not include these diagnostic messages:

```
Matching subject: http://knowledgebooks.com/test#MarkWatsonSoftware
              predicate: http://knowledgebooks.com/test#Customer
```

The `rsesame.jar` file also contains the source code to my Java wrapper code if you want to modify the wrapper interface. If you pass a file name to the `SesameWrapper` constructor, then the in-memory RDF repository is also persisted to a file. If you choose to use a persistent RDF repository, then also use the `close` method on the object `sw` in the last example.

Sesame is my toolkit of choice when writing Semantic Web applications. In the previous section, I showed you how to use Sesame as a networked web service. In this section, you learned how to embed Sesame directly in your JRuby scripts and applications. Whereas Sesame is released under a liberal MIT or BSD-style open source license, you could instead use a commercial product to accomplish a similar result. In the next section, I'll show you how to use one such product that is available with limitations for free use.

Using the AllegroGraph RDF Data Store

The software vendor Franz (`http://franz.com`) sells a very capable RDF data store named AllegroGraph. Franz distributes a no-cost version that you can download and install from `http://agraph.franz.com/allegrograph`. It offers a limit of 50 million triples, which is quite large for the purpose of experimenting with your own data or using small- and medium-size public RDF data sets. There are many public RDF data sets on the Web, and the number of triples is usually listed with each data set.

■**Tip** You can find public RDF data sets listed at `http://en.wikipedia.org/wiki/Linked_Data` and `http://www.rdfabout.com/demo/sec/`.

AllegroGraph, written in Common Lisp, offers client libraries for Lisp, Java, Python, and Ruby. AllegroGraph also supports the Sesame REST APIs that you used in the last section. I'll use the Ruby gem `activerdf_agraph` to directly access an AllegroGraph service. A good additional resource for you to read is Franz's tutorial on using AllegroGraph with Rails: `http://agraph.franz.com/allegrograph/allegrograph-on-rails.lhtml`. This tutorial has instructions for installing `activerdf_agraph` and its dependencies:

```
gem install activerdf activerdf_sparql builder uuidtools
svn co http://activerdfagraph.rubyforge.org/svn/trunk/ agraph
cd agraph
rake package
gem install pkg/activerdf_agraph-*.gem
```

For the rest of this section, I'll assume that you have installed activerdf_agraph, that you've installed the no-cost version of AllegroGraph, and that you've started the server like this:

```
./AllegroGraphServer --http-port 8111
```

As you saw in the previous section on Sesame, the activerdf_agraph library uses the Sesame REST APIs. This library has several dependencies, so you might have some problems installing it—it took me a few tries to get a good installation. If you downloaded the activerdf_ agraph library from RubyForge as I suggested in the last section, you then have the source code and examples in a directory called agraph. To try the following example, change the directory to agraph/examples:

```
myMacBook-2:examples markw$ irb
>> require 'activerdf_agraph'
=> true
>> server = AllegroGraph::Server.new('http://localhost:8111/sesame')
=> #<AllegroGraph::Server:0x193857c @uri="http://localhost:8111/sesame">
>> repo = server.new_repository('test2')
 => #<AllegroGraph::Repository:0x193384c
@uri="http://localhost:8111/sesame/repositories/test2">
 >> repo.load_ntriples('foaf.nt')
=> nil
>>  ConnectionPool.add_data_source(:type => :agraph, :repository => repo)
=> #<AllegroGraph::Adapter:0x1912444 @writes=true,
@repository=#<AllegroGraph::Repository:0x193384c
@uri="http://localhost:8111/sesame/repositories/test2">, readstrue
 >> repo.query(<<EOF)
SELECT ?s WHERE {
  ?s
  <http://www.w3.org/1999/02/22-rdf-syntax-ns#type>
  <http://www.w3.org/2000/01/rdf-schema#Class>
}
EOF
=> "<?xml version="1.0"?>
          <sparql xmlns="http://www.w3.org/2005/sparql-results#">
           <head>
              <variable name="s"/>
           </head>
           <results>
```

```
      <result>
        <binding name="s">
          <uri>http://xmlns.com/foaf/0.1/Agent</uri>
        </binding>
      </result>
      <result>
        <binding name="s">
          <uri>http://xmlns.com/foaf/0.1/Person</uri>
        </binding>
      </result>
      <result>
        <binding name="s">
          <uri>http://xmlns.com/foaf/0.1/Organization</uri>
        </binding>
      </result>
    </results>
  </sparql>"
```

You'll use the `activerdf_agraph` gem again in the section "Loading SEC RDF Data from RdfAbout.com into AllegroGraph," when I show you an example using information from the U.S. Securities and Exchange Commission (SEC).

Testing with Large RDF Data Sets

So far, you have learned how to configure and use three RDF data stores using small test RDF input files. If you'd like to experiment with larger data sets in your projects, you have many options to choose from. By experimenting with larger data sets, you will understand the constraints of memory size, the time it takes to load data, and the time required for both simple and complex SPARQL queries.

I usually prefer and recommend using the Sesame RDF data store because of its easy setup, its multiple inferencing options, its ability to handle large RDF data sets, and its web interface for managing persistent RDF repositories. However, it is also worthwhile for you to experiment with larger RDF data stores using AllegroGraph and Redland. The AllegroGraph commercial product supports scaling repositories to multiple servers and especially benefits the Common Lisp developer. You have already seen how convenient it is to use Redland in embedded Ruby applications, so it is useful to understand its inferencing limitations and general scalability.

Sources of Large RDF Data Sets

There are many public RDF data sets on the web. For example, Joshua Tauberer maintains the web site RdfAbout.com, which provides U.S. SEC data, U.S. census data, and data concerning voting records for members of the U.S. Congress. Another good source of data is DBpedia (http://dbpedia.org), an RDF data set containing information extracted from Wikipedia pages. The RDFizing and Interlinking the EuroStat Data Set Effort (Riese) project is a very good example of a large-scale system that converts data from multiple sources into RDF data and provides interfaces for both human and software agents (see `http://riese.joanneum.at/about.html` for an overview).

There are also public RDF data sets that deal with knowledge representation—that is, they define ontologies of both physical things and ideas or concepts. The Upper Mapping and Binding Exchange Layer (UMBEL) project has taken the upper 20,000 concept classes and their relationships from the OpenCyc knowledge-encoding project and converted this data to RDF (the project web site is `http://www.umbel.org`).

Loading RDF Data from UMBEL into Redland

I downloaded two N3 data files from the `http://www.umbel.org` web site: `umbel.n3` contains class and property definitions, and `umbel_subject_concepts.n3` contains subjects and associated concept definitions. These two files are a small subset of the UMBEL data sets; they contain about a quarter of a million triples.

The example file `src/part2/redland_umbel_data.rb` contains the example program used in this section. You can get the source code from the Apress web site; here, I am just going to show you a few snippets from this source file.

■Note You have already used the N3 RDF format in examples. There is a simpler subset of N3 called "Turtle" that is defined at `http://www.w3.org/2001/sw/DataAccess/df1/`. You can find an RDF primer about using Turtle at `http://www.w3.org/2007/02/turtle/primer/`.

This example program is similar to the previous example `redland_demo.rb`, but here you need to specify that input files use the Turtle syntax:

```
PARSER_RDF_TYPE="turtle"
```

As per the previous example, you can load any number of RDF input files and then perform SPARQL queries:

```
load_rdf_data("file:/Users/markw/Documents/WORK/RDFdata/UMBEL/umbel.n3")
umbel_subject_concepts_small.n3")
 load_rdf_data("
file:/Users/markw/Documents/WORK/RDFdata/UMBEL/umbel_subject_concepts.n3")
 query("SELECT ?s ?p ?o WHERE { ?s ?p ?o }
         LIMIT 5")
query("PREFIX sc: <http://umbel.org/umbel/sc/>
         SELECT ?p ?o WHERE { sc:AcerComputer ?p ?o }
         LIMIT 15 ")
```

Sample output from the second query looks like this:

```
<http://www.w3.org/2002/07/owl#equivalentClass>
<http://sw.opencyc.org/2008/06/10/concept/en/AcerComputer>
```

```
<http://www.w3.org/2004/02/skos/core#definition>
 "The collection of all Acer computers. A type of PersonalComputer. The collection
AcerComputer is a ComputerTypeByBrand and a SpatiallyDisjointObjectType.@en"

<http://www.w3.org/1999/02/22-rdf-syntax-ns#type>
<http://umbel.org/umbel/ac/Computers_Topic>
```

The first result states that sc:AcerComputer is an OWL-equivalent class to an OpenCyc concept instance for Acer Computer. The second result shows the text description, and the third result indicates that sc:AcerComputer is a computer topic. I used a fairly small RDF data set in this example, but in the next two subsections I'll use a much larger data set with Sesame and AllegroGraph.

Loading SEC RDF Data from RdfAbout.com into Sesame

The SEC data contains about 1.6 million triples. I reconfigured my Tomcat and Sesame setup to increase the amount of memory allocated to the Java process by editing tomcat/bin/catalina.sh, adding this line near the top of the file:

```
JAVA_OPTS=-Xmx1200m
```

This increased the Java Virtual Machine (JVM) maximum memory setting to 1.2 gigabytes. You can also pass options on the command line when starting Tomcat with the Sesame web applications. When creating a new Sesame RDF repository, I used the **Native Java Store RDF Schema and Direct Type Hierarchy** option. I then loaded the sec.nt file available at the web site http://www.rdfabout.com. This process took about 30 minutes, but with a persistent RDF data store the repository was immediately available for queries the next time I started Tomcat with Sesame.

You'll be using three URI prefixes when working with the SEC RDF data. The namespace <http://xmlns.com/foaf/0.1/> defines Friend of a Friend (FOAF) relationships such as people's names, relationships between people, and so on. The URIs <http://www.rdfabout.com/rdf/schema/ussec/> and <http://www.rdfabout.com/rdf/usgov/sec/id/> do not reference actual RDF or RDFS definitions; they are simply used as unique base URIs.

I will cover the technical details for using the SPARQL query language in Chapter 6, but for the purposes of testing access to the SEC data, here is a sample SPARQL query that uses three URI prefixes and finds people who have a business relationship with both Apple and Google:

```
PREFIX foaf:  <http://xmlns.com/foaf/0.1/>
PREFIX sec: <http://www.rdfabout.com/rdf/schema/ussec/>
PREFIX seccik: <http://www.rdfabout.com/rdf/usgov/sec/id/>
 SELECT DISTINCT ?name WHERE {
    [foaf:name ?name]
       sec:hasRelation [ sec:corporation [foaf:name "APPLE INC"] ];
       sec:hasRelation [ sec:corporation [foaf:name "Google Inc."] ].
}
```

This SPARQL query returns only two results: LEVINSON ARTHUR D and SCHMIDT ERIC E. The following types are used in this repository and will be useful when you load sec.nt into your local Tomcat/Sesame system and try some interactive queries:

```
foafcorp:Company
foaf:Person
sec:OfficerRelation
sec:DirectorRelation
sec:TenPercentOwnerRelation
```

In the next section, I will show you how to load this same RDF data set into AllegroGraph.

Loading SEC RDF Data from RdfAbout.com into AllegroGraph

For this example, I am assuming that you have installed activerdf_agraph and the no-cost version of AllegroGraph and that you have started the server using this command:

```
./AllegroGraphServer --http-port 8111
```

AllegroGraph does not currently read N3 format files, so use the Redland rapper utility to convert sec.n3 to sec.nt (N-Triple format):

```
rapper -o ntriples -i turtle sec.n3 > sec.nt
```

Use the following code snippet to load the same SEC data from the last section into AllegroGraph and perform the same SPARQL query:

```
require 'activerdf_agraph'
server = AllegroGraph::Server.new('http://localhost:8111/sesame')
repo = server.new_repository('sec2')
repo.load_ntriples('/Users/markw/RDFdata/RdfAbout_SEC_data/sec.nt')
ConnectionPool.add_data_source(:type => :agraph, :repository => repo)
repo.query(<<EOF)
REFIX foaf:   <http://xmlns.com/foaf/0.1/>
PREFIX sec: <http://www.rdfabout.com/rdf/schema/ussec/>
PREFIX seccik: <http://www.rdfabout.com/rdf/usgov/sec/id/>
SELECT DISTINCT ?name WHERE {
    [foaf:name ?name]
        sec:hasRelation [ sec:corporation [foaf:name "APPLE INC"] ];
        sec:hasRelation [ sec:corporation [foaf:name "Google Inc."] ].
}
EOF
```

This example takes a very long time to run and requires about 2.5 gigabytes of resident RAM; the performance issues are caused by the Ruby client code and not AllegroGraph. A better way to load large data sets is to use a server-startup command file (containing Common Lisp code) and start the server using:

```
./AllegroGraphServer --http-port 8111 --http-init-file load.lisp
```

You can then put code to load a repository in this init file. Here is an example load.lisp file for the SEC RDF repository (the path is for my MacBook; you would have to change this depending on where you put the sec.nt file):

```
(in-package :db.agraph.user)
(create-triple-store "/tmp/sec")
(print "*** Starting to load sec.nt")
 (load-ntriples
"/Users/markw/Documents/WORK/RDFdata/RdfAbout_SEC_data/sec.nt" :db "sec")
 (print "*** Done loading sec.nt")
```

As a commercial product, AllegroGraph is a good choice if you are a Common Lisp developer. The combination of an interactive Lisp development environment with either embedded or remote-server use of AllegroGraph is very effective. I find using the AllegroGraph Ruby and Python web-service client libraries to be less convenient, so I prefer using Sesame with Ruby web-service clients.

Using Available Ontologies

When you can find them, it is almost always best to reuse existing public ontologies that support your application domain. A good resource is SchemaWeb (http://www.schemaweb.info), which features ontologies covering many application domains. You might end up designing your own ontology for an application domain, but you should begin by reading what other people working in your domain have already done.

The Dublin Core Metadata project supports making RDF statements about most aspects of publishing information. I have used Dublin Core RDF statements on my main web site since 2002. The namespace abbreviation dc: is usually used for Dublin Core, and you will frequently see it in public RDF data.

You've also used the FOAF ontology for making statements describing people and the relationships between people in the SEC data. Refer to both the site http://www.foaf-project.org and the use of the properties in the FOAF namespace in the SEC RDF data.

The Simple Knowledge Organization System (SKOS) was created by the W3C to make statements about subject headings, classifications, and taxonomies. SKOS uses RDF, so it lacks the formal semantics of OWL ontologies. I think that SKOS's most useful feature is its support for dealing with taxonomies by specifying a hierarchy of type (or class) relationships.

Wrapup

I find Redland to be a good toolkit for embedded RDF stores in Ruby applications when I do not need support for RDFS inference. In Chapter 7, I will implement two networked RDF stores: Sesame REST APIs layered on top of Redland, and JRuby with Sesame and Sinatra. I am going to assume that for the purpose of the examples in later chapters, you'll be using an RDF data store that supports the Sesame REST APIs—that is, either Sesame, AllegroGraph, or my REST interfaces for Redland and Sesame that I'll develop as example applications in Chapter 7.

After working through this chapter, you should be comfortable installing and using at least one of the RDF data stores that I discussed. I'll go into more detail about using SPARQL and reasoning in Chapter 6, so you should choose either Sesame or AllegroGraph for working through those examples. If you choose to use Sesame, then you can save some time by running interactive SPARQL queries using the Workbench web application.

So far, you have been using either simple example RDF data sets or larger data sets that are publicly available on the Web. In Chapter 10, you'll automatically generate RDF data containing information about the relationships between data on two different web sites. You'll continue in Chapter 11 using both publicly available RDF data sources and RDF data that you'll generate in example applications.

■ ■ ■

## CHAPTER 6

Performing SPARQL Queries and Understanding Reasoning

You started to use the SPARQL query language in Chapters 4 and 5. Now you'll dig deeper to learn how queries are performed and how reasoning (also known as *inferencing*) helps us to combine RDF data stores that use different ontologies and to discover implicit relationships in data that are not explicitly stated. Before you look at more practical example queries, I will define some terms that you need to be familiar with and provide a more formal definition of SPARQL query syntax. If you followed along with the SPARQL examples in Chapters 4 and 5, you gained the practical experience necessary to better understand the more formal approach in this chapter.

RDF and SPARQL are key technologies for developing Web 3.0 applications. As needed, I will continue to introduce new techniques for using RDF and SPARQL to help you work through the application examples.

■**Note** I used the Sesame Workbench web application to create and verify the examples in this chapter. I like to keep a separate text editor open so I can paste into the Workbench snippets containing RDF triples and also sample SPARQL queries. I use the text in the text editor as a history of what I am experimenting with when using the web application.

Defining Terms

I have used terms such as *URI*, *triple*, and *literal* without formal definitions, introducing them with simple examples instead. Now is the time to provide concrete definitions.

URI

A URI represents a resource. In most examples so far, I have represented URIs using an N3-notation prefix abbreviation, as in this example where I use a prefix kb: to shorten URIs in the remainder of an RDF document:

```
@prefix kb: <http://knowledgebooks.com/test#> .
kb:Business
```

This is equivalent to using an N-Triple–style absolute URI:

```
<http://knowledgebooks.com/test#Business>
```

The `@prefix` keyword is specific to formats N3 and Turtle (a simplified version of N3). You also use the `PREFIX` keyword in SPARQL queries to define prefix abbreviations. If you need to refer to the RDF document in statements contained in the document, use this notation for the URI of the current document resource: `<>`.

As an example, you might want to annotate an RDF document using these RDF statements inside the document:

```
@prefix dc:<http://purl.org/dc/elements/1.1/> .
<> dc:title "KnowledgeBooks.com Ontologies"
<> dc:creator="Mark Watson"
```

This example uses the Dublin Core Metadata Initiative (`http://dublincore.org`) properties `title` and `creator`. In this example, `<>` is the resource URI for the document containing these triples. Another type of resource is an RDF literal, which I'll describe in the next section.

RDF Literal

An RDF literal is a string that is optionally followed by a nationality suffix:

```
"a string with no country specification"
"a string in English"@en
"a Canadian string literal"@ca
"Bonjour, je suis Française"@fr
```

You can use an RDF literal only as the object part of an RDF triple.

Because more of the web and eventually RDF data will not be in English, it's becoming increasingly important to support and use both internationalization and specification of human-language technologies (not always the same thing). It is possible to add filters to SPARQL queries to selectively match triples based on the nationality of RDF literals.

RDF Typed Literal

An RDF typed literal uses XML data-type designations after a string-literal value. In previous examples, you saw string literals used for values in the object part of a triple. There are standard definitions for integer values, floating-point values, and Boolean values. You tag string values with a data type by following string constants with "`^^xsd:`" and the data type. You can also specify custom types. These examples are self-explanatory:

```
"KnowledgeBooks.com Ontologies"
"Another string literal"^^xsd:string
"1001"^^xsd:integer
1002
"3.14159"^^xsd:decimal
"false"^^xsd:boolean
true
"email:mark@test.com date:02032009"^^<http://knowledgebooks.com#emdate>
```

As you can see in these examples, you can use short forms or express literal values as strings followed by an XML type constant. You have seen SPARQL queries that use a regular-expression filter on a string-literal value in a triple. You can also specify numeric comparison filters in SPARQL queries like this:

```
SELECT ?person ?age
WHERE {
  ?person <http://knowledgebooks.com#age> ?age FILTER (?age > 21)
}
```

■ **Note** We have been using the XML namespace: @prefix xsd (find details at http://www. w3.org/2001/XMLSchema#).

Blank RDF Node

A blank RDF node, or bNode, is an unnamed node that you use to build more complicated data structures that can't be expressed with a single triple value. Blank nodes can appear both inside RDF documents and as parts of SPARQL queries. Blank nodes in a document can be given any name that is only visible in the context of the document. They start with an underscore character or a pair of empty angle brackets. Assuming that the namespaces dc: and kb: are defined with predicates used in the following example, here is a sample use of blank nodes in a small RDF document:

```
@prefix kb:<http://knowledgebooks.com/> .
@prefix dc:<http://purl.org/dc/elements/1.1/> .

<> kb:contains _:base01 .
_:base01 dc:author _:author01 .
_:base01 dc:author _:author02 .
_:author01 kb:name "John Smith" .
_:author02 kb:name "Susan Jones" .
_:base01 dc:title "Small Lake Currents" .
```

A SPARQL query against this example could return a result that John Smith and Susan Jones wrote the paper "Small Lake Currents," but the blank RDF nodes have no scope or meaning outside the document. It is worth looking at a few queries with this last example:

```
SELECT ?s ?o WHERE { ?s dc:title ?o }
```

There is one query result, with the subject equal to `_:node13tbbol8vx4` (an automatically generated unique name) and the object equal to the literal value `Small Lake Currents`. You can also request any number of matching variables for complex (join-style) queries:

```
SELECT ?bn ?bn2 ?name ?title WHERE {
 ?bn dc:title ?title .
 ?bn2 kb:name ?name .
}
```

The output is:

bn	bn2	name	title
_:node13tbbol8vx4	_:node13tbbol8vx5	"John Smith"	"Small Lake Currents"
_:node13tbbol8vx4	_:node13tbbol8vx6	"Susan Jones"	"Small Lake Currents"

The blank-node values are arbitrarily set by the RDF data store. You can use blank nodes as a mechanism for creating complex data structures in RDF that require more than three values. You'll also use them later in `CONSTRUCT` queries.

RDF Triple

I've been using RDF triples and I have defined them informally. A strict definition is that a triple is a data element with three parts:

- *Subject*: An RDF URI reference or a blank node
- *Predicate*: A URI
- *Object*: A URI, literal, RDF literal, or blank node

Some RDF data stores like AllegroGraph augment these three values with an optional graph value. When I use AllegroGraph, I don't use the graph parameter because it hinders portability. Similarly, Sesame augments triples loaded from a URI resource with a field called `context`, which is assigned the value of the source document's URI. This is a nonportable extension.

It is useful to visualize RDF graphs using Graphviz, as you did in Chapter 4 (Figures 4-2, 4-3, and 4-4).

RDF Graph

An RDF graph is a set of RDF triples. So far, I've been discussing a set of triples as all of the triples stored in a specific repository in an RDF data store. The triples comprising the `WHERE` clause of a query also define a graph. Graphs do not have to be completely connected, and in general, an RDF graph will not be completely connected except in simple cases with few nodes.

Comparing SPARQL Query Syntax

You saw SPARQL SELECT queries in Chapters 4 and 5. As you can see in Figure 6-1, there are three additional types of queries: CONSTRUCT, DESCRIBE, and ASK.

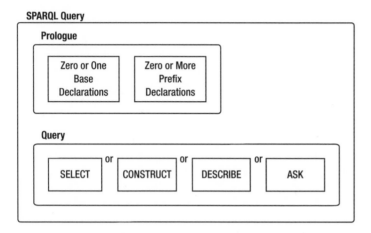

Figure 6-1. *A SPARQL query has optional prologue elements followed by one of four query types.*

You have seen several examples of queries using PREFIX declarations. Here is an example query using a BASE declaration:

```
BASE <http://knowledgebooks.com/>
SELECT ?article_uri ?title ?summary
WHERE {
  { ?article_uri storyType sports } UNION { ?article_uri storyType recreation } .
  ?article_uri title ?title .
  ?article_uri summary ?summary .
}
```

Compare this to the same query that defines a prefix kb: and uses this prefix in the query:

```
PREFIX kb:<http://knowledgebooks.com/>
SELECT ?article_uri ?title ?summary
WHERE {
  { ?article_uri kb:storyType kb:sports } UNION
    { ?article_uri kb:storyType kb:recreation } .
  ?article_uri kb:title ?title .
  ?article_uri kb:summary ?summary .
}
```

Using a BASE declaration allows a shorter and terser query, but my personal preference is to use PREFIX declarations and use the prefixes like kb:title so there is no ambiguity of which namespace a term belongs to. I think this makes queries easier to read and understand. You might see queries that use one BASE declaration and one or more PREFIX declarations.

I'll use the triples in the file src/part2/data/sample_news.n3 in this section (the source code for this chapter is downloadable from the Apress web site). Figure 6-2 shows the triples defined in this file.

Subject	Predicate	Object
kb:oak_creek_flooding	kb:storyType	kb:disaster
kb:oak_creek_flooding	kb:summary	"Oak Creek flooded last week affecting 5 businesses"
kb:oak_creek_flooding	kb:title	"Oak Creek Flood"
kb:bear_mountain_fire	kb:storyType	kb:disaster
kb:bear_mountain_fire	kb:summary	"The fire on Bear Mountain was caused by lightning"
kb:bear_mountain_fire	kb:title	"Bear Mountain Fire"
kb:trout_season	kb:storyType	kb:sports
kb:trout_season	kb:storyType	kb:recreation
kb:trout_season	kb:summary	"Fishing was good the first day of trout season"
kb:trout_season	kb:title	"Trout Season Starts"
kb:jc_basketball	kb:storyType	kb:sports
kb:jc_basketball	kb:summary	"Local JC Basketball team lost by 12 points last night"
kb:jc_basketball	kb:title	"Local JC Lost Last Night"

Figure 6-2. *Triples in the file sample_news.n3 as displayed in the Sesame Workbench*

You can find the examples in the next sections as text files in the directory src/part2/data, with file extensions of .txt. (The RDF data files in this directory have N3, NT, or OWL extensions: .n3, .nt, or .owl.) These text files contain many more examples than the few that I chose to show in the following text, so you might want to try them out.

SPARQL SELECT Queries

The most common SPARQL query type that you are likely to use is the SELECT query, so you'll see several useful examples in this section. The first example involves the UNION keyword, which you use to specify alternative graph-matching patterns in the WHERE clause of queries (reformatted to fit line width):

```
PREFIX kb:<http://knowledgebooks.com/>
SELECT DISTINCT ?article_uri ?title ?summary
WHERE {
  { ?article_uri kb:storyType kb:sports }
      UNION
  { ?article_uri kb:storyType kb:recreation } .
  ?article_uri kb:title ?title .
```

```
     ?article_uri kb:summary ?summary .
}
```

Note that you can always surround any triple subgraph pattern with curly brackets; in this case, I used curly brackets to clearly indicate the subgraph patterns immediately before and immediately after the UNION keyword.

In the preceding query, I specified a list of three variables to return values for: ?article_uri, ?title, and ?summary. If you want a query to return values for all variables appearing in the WHERE clause, then use an asterisk character as an argument for SELECT:

```
SELECT DISTINCT *
```

You'll see an alternative approach of using a FILTER instead of a UNION operation in the next section. Using a FILTER is more general, but queries that use a UNION are easier to read and understand when you want to match with one of two or three alternatives. In this example, I used UNION to allow matches with different predicates, but you can also use UNION for alternative subject and object parts of triples. Figure 6-3 shows the results for the last query, as seen in the Sesame Workbench. I will not always show query results in the rest of this chapter because I am assuming that you are trying the queries yourself. The queries for this section (with many more examples) are in the file src/part2/data/sparql_select_test.txt.

Article_uri	Title	Summary
kb:trout_season	"Trout Season Starts"	"Fishing was good the first day of trout season"
kb:jc_basketball	"Local JC Lost Last Night"	"Local JC Basketball team lost by 12 points last night"

Figure 6-3. *SPARQL SELECT query results*

The next example uses the OPTIONAL keyword to specify a WHERE clause with subgraphs that do not have to be matched. The news-article triples we are using all have both a title and a summary defined for each article URI. Here, I add a new article without a summary:

```
@prefix kb: <http://knowledgebooks.com/> .

kb:jc_bowling kb:storyType kb:sports ;
              kb:title "JC Bowling Team to Open Season" .
```

The following query uses the OPTIONAL keyword:

```
PREFIX kb:<http://knowledgebooks.com/>

SELECT DISTINCT ?title ?summary
WHERE { ?article_uri kb:title  ?title .
        OPTIONAL { ?article_uri kb:summary ?summary } }
}
```

The query result, which matches the two new triples I just added, is shown in Figure 6-4.

Title	Summary
"Oak Creek Flood"	"Oak Creek flooded last week affecting 5 businesses"
"Bear Mountain Fire"	"The fire on Bear Mountain was caused by lightning"
"Trout Season Starts"	"Fishing was good the first day of trout season"
"Local JC Lost Last Night"	"Local JC Basketball team lost by 12 points last night"
"JC Bowling Team to Open Season"	

Figure 6-4. *A query using the OPTIONAL keyword allows match variables to remain undefined.*

In Figure 6-4, the last query result does not have a value assigned to the ?summary variable. You can add constraints to optional matching subgraphs:

```
PREFIX kb:<http://knowledgebooks.com/>
SELECT DISTINCT ?title ?summary ?page_count
WHERE {
    ?article_uri kb:title  ?title .
    OPTIONAL { ?article_uri kb:summary ?summary } .
    OPTIONAL { ?article_uri kb:page_count ?page_count . FILTER (?page_count > 1) } .
}
ORDER BY ?title
LIMIT 20
OFFSET 20
```

In this example, the match variable ?page_count is set only if there is a triple matching the subgraph { ?article_uri kb:page_count ?page_count } and if the value set for the object match variable ?page_count has a numeric value greater than 1. The results are sorted alphabetically by title. If you wanted to sort in descending order instead, then you would use ORDER BY DESC(?title). This query limits the number of returned results to 20. The clause OFFSET 20 indicates that the code should ignore the first 20 matches when returning results.

You can use the sameTerm operator to see if values matched to different variables are the same:

```
PREFIX kb:<http://knowledgebooks.com/>
SELECT DISTINCT ?article_uri1 ?article_uri2 ?predicate1 ?predicate2
WHERE {
    ?article_uri1 ?predicate1 ?o1 .
    ?article_uri2 ?predicate2 ?o2 .
    FILTER (!sameTerm(?article_uri1, ?article_uri2) &&
             sameTerm(?predicate1, ?predicate2)) .
    FILTER regex(?o1, "Season") .
    FILTER regex(?o2, "Season") .
}
```

Figure 6-5 shows the results of this query.

Article_uri1	Article_uri2	Predicate1	Predicate2
kb:trout_season	kb:jc_bowling	kb:title	kb:title
kb:jc_bowling	kb:trout_season	kb:title	kb:title

Figure 6-5. *Results from the example query that used the sameTerm operator*

The SPARQL recommendation (http://www.w3.org/TR/rdf-sparql-query) lists 13 filter operators like sameTerm. For example, you use the operator isIRI to test if a matched value is an IRI.

■**Note** IRI stands for Internationalized Resource Identifier. I have been using the acronyms URI and IRI interchangeably, but they are not quite the same: IRIs use a universal ISO character set.

Another useful operator is isLiteral, which checks whether a matched value is any form of literal. You have already seen several examples that use the filter operator regex for regular-expression matching in literal values.

SPARQL CONSTRUCT Queries

CONSTRUCT queries, which provide a template for creating new triples, are very different from SELECT queries. A SELECT query returns a graph containing copies (with blank nodes replaced by new blank-node values) of triples that match the query patterns. SELECT queries can also return partial triples—just matched subjects and objects, for example. You use CONSTRUCT queries to create new complete triples containing a subject, predicate, and object that are "sliced and diced" from triples matching query patterns. You can construct new triples making statements with different property URIs, reversing the subject and object, and so on.

Templates can use variables that have values assigned while matching the WHERE clause of a CONSTRUCT query:

```
PREFIX mw: <http://markwatson.com/rdf/>
PREFIX kb: <http://knowledgebooks.com/>
CONSTRUCT { ?x mw:synopsis ?o }
WHERE { ?x kb:summary ?o }
```

A template follows the CONSTRUCT keyword in queries. In this example, the variable ?x matches a document URI and the variable ?o matches a string literal for a news-article summary. This template substitutes the predicate mw:synopsis for kb:summary in the original matched triples. The constructed triples are not added to the repository; they are returned as values from the query as a single graph. Triple ordering in the returned graph is immaterial because a graph is a set (an unordered collection with no duplicate elements) of triples.

This first example is simple, but it demonstrates how you use CONSTRUCT queries. One practical use of CONSTRUCT queries is to make local copies of portions of remote repositories and merge them in a new local repository. For instance, you could extend this first example to map a predicate used in a remote repository to a predicate that you use in your application.

Later in this chapter I'll show you another method for merging data using different schemas (see the section "Combining RDF Repositories That Use Different Schemas"). Specifically, you'll use RDFS and OWL inferencing to use data defined with different schemas without converting the data. Here is another CONSTRUCT query example that constructs two types of triples. One triple is a statement that a story type is the subject of an article URI, and the other triple is a statement that two stories have the same story type:

```
PREFIX kb:<http://knowledgebooks.com/>
CONSTRUCT { ?story_type_value kb:subject_of ?article1 .
                  ?article2 kb:same_topic ?article1 .}
WHERE {
  ?article1 ?story_type ?story_type_value .
  ?article2 ?story_type ?story_type_value .
  FILTER ((?article1 != ?article2) &&
          ((?story_type_value = kb:sports) || (?story_type_value = kb:recreation)))
}
```

The filter statements in this query make sure that two matched articles are not the same and that only sports and recreation story types are matched.

In addition to creating new triples using different property names and types, the CONSTRUCT query is useful for constructing more complex subgraphs from several information sources. Conversely, you can use the CONSTRUCT query to convert complex subgraphs with one subject and many predicates and objects to simple subgraphs for new applications. If you think of triples in a repository as a raw material, then CONSTRUCT queries allow you to mold data into new forms and for new purposes.

SPARQL DESCRIBE Queries

A DESCRIBE query can identify one or more URIs used in an RDF repository and then return all triples that have the same subject. For example, this query returns 20 triples, all containing the subject URI matched in the WHERE clause:

```
PREFIX kb:<http://knowledgebooks.com/>

DESCRIBE ?article_uri
WHERE { ?article_uri kb:title "Trout Season Starts" }
```

Multiple URIs can be matched and described in a single DESCRIBE query. For example, this query returns a graph containing 30 triples that have either kb:trout_season or kb:title as a subject:

```
PREFIX kb:<http://knowledgebooks.com/>

DESCRIBE ?article_uri ?predicate
WHERE { ?article_uri ?predicate "Trout Season Starts" }
```

You can use the DESCRIBE query as a starting point for exploring a repository when you want to find all triples with a specific subject URI.

SPARQL ASK Queries

SPARQL ASK queries try to match a subgraph defined by WHERE clauses to a repository graph. They return "yes" if the subgraph is found and "no" if no subgraph is found. For example, here is an ASK query that returns "yes" (assuming that the data in sample_news.n3 is loaded):

```
PREFIX kb:<http://knowledgebooks.com/>
ASK
WHERE { ?article_uri ?predicate "Trout Season Starts" }
```

This query returns a "yes" (or true) value because the subgraph pattern matches two triples. You can verify this for yourself by running this query as a SELECT query. Here is a query with a subgraph that is not found in the repository, so the query result is "no" (or false):

```
PREFIX kb:<http://knowledgebooks.com/>
ASK
WHERE { ?article_uri kb:copyright ?copyright_value }
```

The query result is "no" because there are no triples in the repository that contain the predicate kb:copyright.

Implementing Reasoning and Inference

While you probably have extensive experience using relational databases, I suspect that you might not have deep knowledge of the underlying relational calculus or relational algebra that defines how SQL database queries match rows in database tables and produce results. You could make the same argument about the underlying reasoning (or inference) system for SPARQL queries—that you don't need to know the core mechanics in order to use them effectively. However, knowledge of how inferencing works can help you better use Semantic Web technologies.

The field of artificial intelligence (AI) includes the broad topic of knowledge representation for automated processing. You can use RDF data storage and logic to represent knowledge for software agents to work with. There are opposing constraints in the selection of a logic for RDF reasoning: you want a logic language that supports class inference, determines class membership, infers data relationships that are not explicitly specified, and so on. You also want the inferencing computations to run in a reasonable amount of time for large data sets.

You have a wide range of choices for the level of inference support. Redland is very efficient for storing and retrieving RDF triples, but offers limited inferencing. Some RDF repositories offer more support for RDFS inferencing over class membership and properties. The new OWL standard, which offers a rich data model built on top of RDF and RDFS, features the most flexible inferencing support for Semantic Web applications. It provides this support in the form of three types of reasoning: OWL Light, OWL DL (named for its correspondence with description logics), and OWL Full. The commercial AllegroGraph product supports a compromise between RDFS and OWL inferencing that is known as RDFS++. Sesame versions 2.2.2 and above support RDF and RDFS reasoning plus some degree of OWL class reasoning. For OWL DL support, you can use version 2 of the Pellet reasoner, which I mentioned in Chapter 4 (available at `http://clarkparsia.com/pellet`). Pellet is released under the GPL.

There are two general approaches to implementing inferencing. Implementing inferencing is probably of most interest to developers of RDF data stores, but it is worth knowing that some data stores like Sesame use forward-chaining inference rules to add additional triples automatically as triples are loaded into the repository. Here's an example: if a property `isBrother` is a subproperty of `isSibling`, and if you explicitly add a triple to Sesame stating that Ron is a brother, then a type-inference rule can automatically generate an additional triple stating that Ron is a sibling. An alternative approach, which AllegroGraph uses, is performing inferencing during queries and not permanently adding new implicit (that is, inferred) triples to the repository.

I have been using Sesame for most of examples in this book because its BSD-style license allows its use in commercial projects. Plus, Sesame provides good support for what you need when getting started on Semantic Web projects: an efficient RDF data store, RDF and RDFS inference support, and a convenient administration web application. I'll assume the use of Sesame for the following discussion, but this discussion is valid for any RDF data store that supports at least RDFS inference. When you create a new repository using the Sesame Workbench, you select the level of inference support that you require (see Figure 6-6).

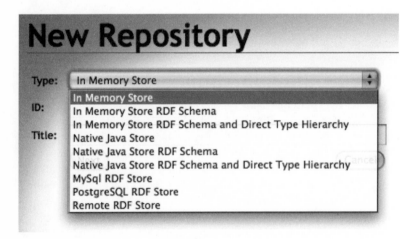

Figure 6-6. *Options for Sesame RDF data stores*

The file `src/part2/data/inferencing_1.txt` contains all of the examples in this section. I am using a data store with "RDF Schema and Direct Type Hierarchy" support to develop and test the examples in this section.

RDFS Inferencing: Type Propagation Rule for Class Inheritance

To understand RDFS inferencing, you need to understand type propagation. The concept of type propagation is simple: if class A is a subclass of B, and if x is a member of class A, then x is also a member of class B. I'll demonstrate type propagation with a short example that starts with loading the following data into a fresh empty repository:

```
@prefix rdf: <http://www.w3.org/1999/02/22-rdf-syntax-ns#> .
@prefix rdfs: <http://www.w3.org/2000/01/rdf-schema#> .
@prefix kb: <http://knowledgebooks.com/> .
@prefix person: <http://knowledgebooks.com/person/> .

kb:Sibling rdfs:subClassOf rdfs:Class .
kb:Brother rdfs:subClassOf kb:Sibling .
 person:mark rdf:type kb:Brother .
```

This snippet states that the class Sibling is a subclass of the global class Class, that class Brother is a subclass of class Sibling, and that person:mark is of type kb:Brother (that is, person:mark is an instance of class kb:Brother). You would expect the following query to return a value of person:mark to match the subject part of the WHERE clause subgraph:

```
SELECT DISTINCT ?s
WHERE { ?s rdf:type kb:Brother }
```

If you are following along with this example using Sesame Workbench, you see that there is one result with the match variable ?s set to person:mark. Because you are using a repository that supports RDFS inferencing, the following query returns the same result because kb:Brother is a subclass of kb:Sibling:

```
SELECT DISTINCT ?s
WHERE { ?s rdf:type kb:Sibling }
```

You will see in the next section that the type-propagation rule also works for properties, with examples using properties parent, mother, and father.

RDFS Inferencing: Type Propagation Rule for Property Inheritance

Perhaps even more useful than defining new classes is defining new application-specific *predicates* (which I'll also refer to as *properties*). Class membership is a statement about the type of resource, and properties (or predicates) define relationships between resources. In the example in the last section, kb:Brother and kb:Sibling are class definitions—note that class names are capitalized. You'll now look at a similar example that defines some new properties (note that property names start with a lowercase letter). Start by loading the following triples into a fresh (empty) new repository:

```
@prefix rdf: <http://www.w3.org/1999/02/22-rdf-syntax-ns#> .
@prefix rdfs: <http://www.w3.org/2000/01/rdf-schema#> .
@prefix kb: <http://knowledgebooks.com/> .
@prefix person: <http://knowledgebooks.com/person/> .

kb:mother rdfs:subPropertyOf kb:parent .
kb:father rdfs:subPropertyOf kb:parent .

person:ron kb:father person:anthony .
```

Because `kb:father` is defined as an `rdfs:subPropertyOf kb:parent`, both of the following queries return a single result with ?s bound to `person:ron` and ?o bound to `person:anthony`:

```
SELECT DISTINCT ?s ?o
WHERE { ?s kb:father ?o }

SELECT DISTINCT ?s ?o
WHERE { ?s kb:parent ?o }
```

Both `rdfs:subClassOf` and `rdfs:subPropertyOf` are transitive: if A is a subclass of B and B is a subclass of C, and if x is a member of class A, then x is also a member of classes B and C. Likewise, if property p1 is a subproperty of p2 and p2 is a subproperty of p3, then if x p1 y is a valid statement, then x p2 y and x p3 y are also valid statements.

Using rdfs:range and rdfs:domain to Infer Triples

In a practical sense, you already know everything necessary to effectively model RDF data using RDFS and to effectively use SPARQL to use your models and data. However, there is one other useful modeling construct that I will discuss before closing out this chapter: using `rdfs:range` and `rdfs:domain` to effectively add inferred triples to a repository.

You should understand that range and domain restrictions will not cause errors when you load data. Instead, the RDF data store will infer new triples that make statements about the types of subject and object resources. The next example clarifies this. The range of a predicate (or property) indicates the class (or type) of a matching object in a triple. The domain of a predicate indicates the class of a matching subject in a triple. As an example, you can specify that subjects of triples with the predicate `kb:mother` must be of class `kb:Female`, and subjects of triples with the predicate `kb:father` must be of class `kb:Male`:

```
kb:mother rdfs:domain kb:Female .
kb:father rdfs:domain kb:Male .

person:kate rdf:type kb:Female .
person:kate kb:father person:bill .
```

This example is important. You might expect an error to be thrown because `person:kate` is of type `kb:Female` and the `rdfs:domain` of the `kb:father` predicate is `kb:Male`. But no error is thrown, and instead the inferencing mechanism provided by the RDF data store infers a new triple stating that `person:kate` is of type `kb:Male`. The repository now contains triples stating that `kate` is both a male and a female. You might question how this type inference can be useful. To answer this, I'll show you a better example where type information is not over-specified:

```
kb:mother rdfs:domain kb:Female .
kb:father rdfs:domain kb:Male .

person:mary kb:mother person:susan .
```

When I load these three statements (with required prefix statements) into a fresh repository, a new triple that Mary is a female is inferred. I assigned the context to the resource URI `<file://infer6.n3>` to these three statements when I loaded them. When you explore this context, you see the expected triples in Figure 6-7.

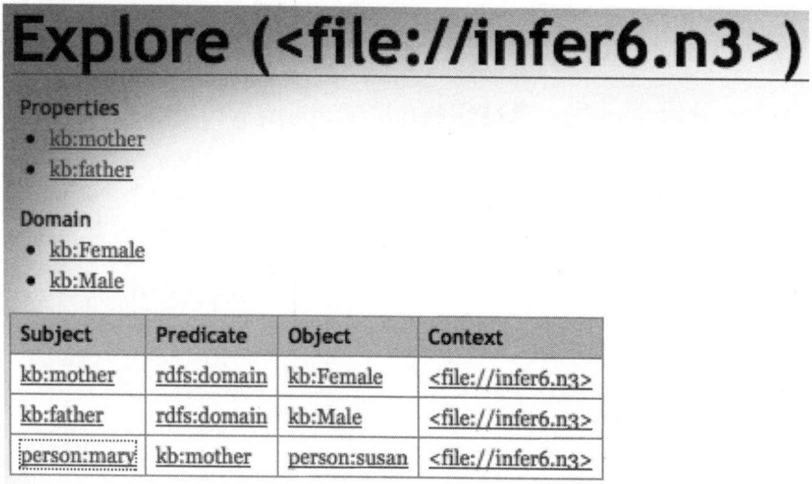

Explore (<file://infer6.n3>)

Properties
- kb:mother
- kb:father

Domain
- kb:Female
- kb:Male

Subject	Predicate	Object	Context
kb:mother	rdfs:domain	kb:Female	<file://infer6.n3>
kb:father	rdfs:domain	kb:Male	<file://infer6.n3>
person:mary	kb:mother	person:susan	<file://infer6.n3>

Figure 6-7. *The view when you explore the context after loading three triples in a new repository*

What you do not see in Figure 6-7 is the inferred triples because exploring a context shows only explicitly defined triples, not inferred triples. However, if I select the link `person:mary` in Figure 6-7, then I see the inferred triples as shown in Figure 6-8. The inferred triples have a blank context value.

Explore (person:mary)

Subject	Predicate	Object	Context
person:mary	rdf:type	rdfs:Resource	
person:mary	rdf:type	kb:Female	
person:mary	kb:mother	person:susan	
person:mary	kb:mother	person:susan	<file://infer6.n3>
person:mary	<http://www.openrdf.org/schema/sesame#directType>	kb:Female	

Figure 6-8. *Inferred triples have a blank context value.*

Sesame applies type-propagation and type-inference rules to all triples as they are loading into a repository, and there is no check for creating new triples that match the subject, predicate, and object values of other triples already in the repository. To combat this duplication, you can use the DISTINCT keyword when performing SELECT queries.

This last example showed how the RDF data store inferred that person:mary is of type kb:Female. A query asking for all females would produce "Mary" because Mary is a mother, even though we never explicitly stated that Mary is a female.

Combining RDF Repositories That Use Different Schemas

People and organizations who publish information on the Web have the freedom to format data for their own uses. As a result, you will find it necessary to merge RDF data from different sources and often deal with different schemas and namespaces. One approach is to use SPARQL CONSTRUCT queries to convert data to a common schema and store it separately. Another approach is to make statements that different classes or different properties are equivalent. Because you have already used CONSTRUCT queries, I will use the second approach here.

If you are using an RDF data store that supports OWL reasoning, then you can use the property owl:sameAs to make statements about class or property equivalency. For example, assume that you have triples modeling data for news articles, but that these triples are in two disjoint namespaces and perhaps use different property names:

```
@prefix kb: <http://knowledgebooks.com/> .
@prefix testnews: <http://testnews.com/> .
@prefix owl: <http://www.w3.org/2002/07/owl#> .

kb:title owl:sameAs testnews:title .

kb:oak_creek_flooding kb:storyType kb:disaster ;
    kb:summary "Oak Creek flooded last week affecting 5 businesses" ;
    kb:title "Oak Creek Flood" .
```

The following two queries provide the same result if OWL reasoning is supported:

```
SELECT DISTINCT ?article_uri1 ?object
WHERE {
    ?article_uri1 kb:title ?object .
}

# works if OWL inferencing is supported:
SELECT DISTINCT ?article_uri1 ?object
WHERE {
    ?article_uri1 testnews:title ?object .
}
```

If OWL reasoning is not supported, then the second query returns no results. When OWL inferencing is not available, you can get the same effect with some extra work in RDFS using the rdfs:subPropertyOf in two statements:

```
kb:title rdfs:subPropertyOf testnews:title .
testnews:title rdfs:subPropertyOf kb:title .
```

Similarly, you can use rdfs:subClassOf in two statements to declare equivalence of two classes.

Wrapup

I concentrated on showing you the most practical aspects of SPARQL queries and inference. You saw one example in the last section showing that the OWL modeling language and description logic are more powerful tools than RDF, RDFS, and RDFS inferencing. The future widespread use of OWL is probable, but RDF, RDFS, and RDFS technologies are easier to implement and use. Because of this, I suspect that they will form the foundation of the Semantic Web and Web 3.0 for the near future.

It is worthwhile mentioning some of the OWL extensions to RDFS. OWL supports making statements about the cardinality of class membership, such as "a child has at most one mother." OWL also supports statements about set union and intersection. The owl:sameAs property that you saw in the last section is used to make statements about the equivalency of two individual resources specified by URIs—in the example I showed you, the individual resources were properties but they could also be classes. You can use the property owl?COL?differentFrom to state that two resources are not the same thing. OWL also supports enumerated types, the ability to make statements on transitivity and symmetry, and the definition of functional properties. OWL is built on the foundation of RDF and RDFS. Both OWL and RDFS are expressed as RDF statements.

In the next chapter, I will design and implement two SPARQL endpoint web portals using both Sesame and Redland as back ends.

CHAPTER 7

■■■

Implementing SPARQL Endpoint Web Portals

A general theme in this book is building Web 3.0 systems in light and flexible ways using the Ruby scripting language combined with existing libraries and applications. I believe that as more "smart," semantically rich data becomes available, the economy-of-scale arguments in favor of using very large systems as opposed to smaller systems diminish because accessing data to drive smaller web portals becomes easier. As it becomes technically easier to federate smaller systems, I expect to see more special-purpose niche applications that interoperate. Combined with an aggressive use of open source software to reduce development costs, smart data sources will make the development, deployment, and maintenance of many small inter-connected systems economically feasible. The examples in this chapter demonstrate simple ways to publish semantically rich data.

You used the Sesame RDF data store and web portal in Chapter 5. While Sesame provides a full-featured web interface and web-service APIs, for some applications a simpler frame-work will suffice. One of the examples I'll develop in this chapter is a lightweight SPARQL endpoint server that uses JRuby, the Sesame RDF data store, and Sinatra. Sinatra is a simple web framework that I find useful for writing web portals. Described by its developers as a domain-specific language (DSL) for developing web applications, Sinatra is an effective tool for building small web-service or web-portal programs (see `http://www.sinatrarb.com` for details).

Another example program I'll build is a SPARQL endpoint using Ruby 1.8.6, Redland, and Sinatra. These examples will have a common front end but different back ends. I will also extend the programs to accept real-time data updates and to provide web services for RDF data-visualization graphs using Graphviz.

With its bundled web front ends, Sesame is full-featured, but I am still motivated to write "hacker"-friendly SPARQL endpoint servers in Ruby. Instead of supplying an admin web inter-face like Sesame, these example web portals read all RDF data files in the directory `rdf_data` and load them into a back-end repository that supports SPARQL queries through a REST-style web-service interface. I hope that you find these implementations both educational and useful for projects requiring a flexible and easily modified implementation in Ruby.

Figure 7-1 shows a UML class diagram for the example web portals that I'll develop in this chapter.

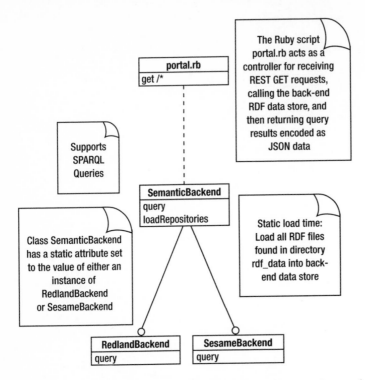

Figure 7-1. *UML class diagram for two SPARQL endpoint web portals*

Even though I am using Sinatra in this chapter, a lot of what we do here will be reusable in Chapter 15 when I combine some material from Parts 3 and 4 into a large example system based on Ruby on Rails.

I package the examples in the directory src/part2/sparql_endpoint_web_service (you can download the source code for this chapter from the Apress web site at http://www.apress.com). The scripts to run the three versions of the portal are in the files portal.rb, portal_auto_load.rb, and portal_graphviz.rb. We will look at the code for these later. I like to package Ruby libraries that I write as gems, and going through the gem-creation process is worthwhile. I use the hoe gem, which you can install using gem install hoe. This puts the command-line utility sow on your execution path, so you can create a new blank gem project using:

```
sow sparql_endpoint_web_service
```

The generated Rakefile includes the hoe gem and adds several rake tasks (use rake -T to view all available tasks). When developing a gem, you'll use these tasks most often:

```
rake clean              # Clean up all the extras
rake clobber_package    # Remove package products
rake debug_gem          # Show information about the gem
rake gem                # Build the gem file sparql_endpoint_web_service-1.0.0.gem
rake package            # Build all the packages
```

```
rake repackage          # Force a rebuild of the package files
rake test               # Run the test suite
```

I will discuss the JRuby/Sesame and the Ruby 1.8.6/Redland back-end implementations later, but for now I'll describe a scheme for supporting two different back ends in one gem library. Figure 7-2 shows the directory structure and files for the gem-library implementation. As you saw in Chapter 5, the file rsesame.jar contains my packaging and a Ruby wrapper for everything that the JRuby/Sesame back end needs in order to use the Sesame Java libraries. The Redland back end requires the installation of both the Redland C libraries and the Ruby rdf/redland gem (see Chapter 5 for reference).

Figure 7-2. *Library directory structure for the gem*

The file sparql_endpoint_web_service.rb is the top-level library file that first tries to load JRuby Java library support—if this operation succeeds, then the Sesame back end is loaded. If there is an error, then I assume we are running Ruby 1.8.6 and I try to load the Redland back end. This file starts by adding the lib subdirectory to the Ruby load path $: and then requires Sinatra and JSON gems. The next three Ruby require statements load three source files from the lib/sparql_endpoint_web_service directory:

```
$:.unshift(File.dirname(__FILE__))

require 'sinatra'
require 'json'

require File.dirname(__FILE__) + '/sparql_endpoint_web_service/semantic_backend.rb'

class SparqlEndpointWebService
  VERSION = '1.0.0'
end
```

The Ruby source file lib/sparql_endpoint_web_service/semantic_backend.rb provides a uniform query interface for back ends and does the work of loading the correct back end depending on whether you are using JRuby or Ruby 1.8.4. The following code defines the class SemanticBackend and initializes it differently for Ruby 1.8.6 or JRuby:

```
class SemanticBackend
  def query a_query
    @@back_end.query(a_query)
  end
  def SemanticBackend.set_back_end be
```

```
      @@back_end = be
    end
    def SemanticBackend.initialized?
      @@back_end != nil
    end
end


  # Test to see if we are running under JRuby or CRuby:
begin
  include Java
  require 'sparql_endpoint_web_service/sesame_backend'
  SesameBackend.loadRepositories
  SemanticBackend.set_back_end(SesameBackend.new)
rescue
  # failed to load Java, so assume that we are running Ruby 1.8.6 and Redland:
  require 'sparql_endpoint_web_service/redland_backend'
  RedlandBackend.loadRepositories
  SemanticBackend.set_back_end(RedlandBackend.new)
end

raise "Could not load an RDF backend" if ! SemanticBackend.initialized?
```

The class SemanticBackend with the method query provides a common interface to the Sesame and Redland back ends. In the next three sections, we will look at the implementation of a common web-portal front end, the implementation of the Sesame back end, and the implementation of the Redland back end.

Checking at runtime whether JRuby or CRuby is in use is a good technique for supporting multiple back-end options. CRuby is the standard Ruby that is implemented in the C programming language. You might prefer using different class hierarchies with common code implemented as a mix-in.

Because of JRuby's excellent runtime support using the Java platform, you might choose to simply strip the CRuby and Redland support from these example portals if you will only be using a Java runtime environment. For my own development style, I like to support both CRuby and JRuby because CRuby has a much faster startup time that makes development faster and more enjoyable. For much of my work, I want to be able to develop with CRuby and then deploy with JRuby.

Designing and Implementing a Common Front End

I am using the lightweight Sinatra library to implement a common front-end web application that accepts REST-style SPARQL queries and returns a JSON payload containing the SPARQL query results.

I am supporting both Ruby 1.8.6 and JRuby in this example. The only difference in setup is that you should install the pure Ruby version of the JSON gem for JRuby and the normal native code JSON gem for Ruby 1.8.6. So install one of these:

```
gem install json_pure    # for JRuby
gem install json         # for Ruby 1.8.6
```

In general, if you want to use Sinatra to write general-purpose web sites, you will want to install extra dependencies for this:

```
gem install rack sinatra builder haml
```

For our purposes in this chapter, you need only the Rack, Builder, and Sinatra gems installed. The following code snippets show the complete script portal.rb, with interspersed explanations. In addition to the Sinatra and JSON gems, I need to load the CGI gem for decoding URL-encoded strings (part of the REST GET HTTP request). I also require the sparql_endpoint_web_service gem from the local library subdirectory:

```
require 'rubygems'
require 'sinatra'
require 'json'
require 'cgi'
require 'lib/sparql_endpoint_web_service'
```

The Sinatra gem internally loads both Rack and Builder. When you require (or load) the Sinatra gem, an internal run method is called after the gem is initialized. This run method detects a Rack installation (pure Ruby, running with Apache, nginx, or the like) and prepares to handle incoming requests.

To make development faster, Sinatra effectively reloads the top-level script for each new request when running in "development mode." This does not happen in "production mode." The Sinatra web site features *The Sinatra Book* (http://www.sinatrarb.com/book.html), which documents how to deploy in production mode and how to use different front-end web servers with Rack. I do not cover deployment in this chapter, but you can refer to the online Sinatra documentation.

The Sinatra DSL defines methods get, put, post, and delete that all work in a similar way, but are responsible for HTTP GET, PUT, POST, and DELETE requests. The following call to the utility method get defines a URL route-matching pattern and associates this pattern with the code defined inside the get call:

```
get '/*' do
  content_type 'text/json'
  if params['query']
    query = CGI.unescape(params['query'])
    ses = SemanticBackend.new
    response = ses.query(query)
    JSON.generate(response)
  else
    ''
  end
end
```

This is the entire top-level Sinatra script—just 16 lines of Ruby code. The pattern /* matches any incoming URL pattern and copies any request variable names and values to the params hash table. The only assumption that I make is that the client sets a variable query to a URL-encoded string value that, when decoded, represents a valid SPARQL query. The method call to content_type sets the returned HTML-header content type for JSON text data. I am using the SemanticBackend class described in the last section to pass the decoded SPARQL query to the back-end RDF data store. The response from the query method call will be an array of results, each of which is an array of string values. The call to JSON.generate encodes this data as JSON text data. If the request does not contain a query, then an empty string is returned to the client.

It is worth looking at an example query and how it is encoded and passed through this Sinatra-based script to the back end. The following code snippet encodes a query string and prints it:

```ruby
require 'cgi'
sample = " SELECT ?s ?o WHERE { ?s <http:://test.com/ontology/#summary> ?o } "
puts sample
puts CGI.escape(sample)
```

The two lines of output are (note that the second line wraps):

```
SELECT ?s ?o WHERE { ?s <http:://knowledgebooks.com/ontology/#summary> ?o }
SELECT+%3Fs+%3Fo+WHERE+%7B+%3Fs+%3Chttp%3A%3A%2F%2Ftest.com%2Fontology%2F➥
%23summary%3E+%3Fo+%7D
```

If you use this web portal with Ruby clients, you will also use the CGI gem to encode SPARQL query strings. I'll show you sample web-service clients later, after implementing Sesame- and Redland-based back ends in the next two sections.

The brevity and readability of the portal.rb script shows the power of using well-designed DSLs such as Sinatra, a custom language built to solve problems in a narrow application domain. Whenever I use a Ruby gem that someone else wrote, I like to start by running some examples to make sure that it meets my current requirements. Before seriously using any gem, I also make a habit of opening the gem's source code in an editor or Ruby IDE and reading through it, both to understand the implementation and also to learn new Ruby coding tricks from the gem's author.

In the next two sections, I'll show you how to implement the two back ends based on Sesame and Redland.

Designing and Implementing the JRuby and Sesame Back End

You previously used Sesame in embedded mode in Chapter 5. Here you'll use the same Java JAR file rsesame.jar, which defines the utility class SesameWrapper and contains all of the runtime libraries required for embedding Sesame in either Java or JRuby applications. In this example, I start by defining a callback-handling class whose method triple is called for each query result. This method accumulates results in the instance attribute @results:

```
require 'lib/rsesame.jar'
include_class "SesameWrapper"

# Use duck typing: define a class implementing the method "triple" :
class SesameCallback
  attr_reader :results
  def initialize
    @results = []
  end
  def triple result_list # called for each SPARQL query result
    ret = []
    result_list.each {|result|
      if result.index('http:') || result.index('https:')
        ret << "<" + result + ">"
      else
        ret << result
      end
    }
    @results << ret
  end
end
```

An instance of class SesameBackend can be used by the class SemanticBackend to provide RDF storage and SPARQL query support. By default, SesameBackend reads all RDF N-Triple files in the subdirectory rdf_data, but you can override the data-directory location by setting the RDF_DATA environment variable to a different location:

```
class SesameBackend
  def SesameBackend.loadRepositories
    if ENV['RDF_DATA']
      rdf_path = ENV['RDF_DATA']
    else
      rdf_path = 'rdf_data'
    end
    @@tsm = SesameWrapper.new
    Dir.entries(rdf_path).each {|fname|
      if fname.index('.nt')
        begin
          puts "* loading RDF repository into Sesame: #{rdf_path + '/' + fname}"
          @@tsm.load_ntriples(rdf_path + '/' + fname)
        rescue
          puts "Error: #{$!}"
        end
      end
    }
  end
  def query sparql_query
    callback = SesameCallback.new
```

```
    @@tsm.query(sparql_query, callback)
    callback.results
  end
end
```

I used the Sesame wrapper-class method `load_ntriples` to process any files ending with the file extension ".nt," but you can also choose from other file-loading methods in the wrapper class: `load_n3` and `load_rdfxml`. The method `sparql_query` instantiates a new instance of the `SesameCallback` class, passes this instance to the Sesame wrapper-class method `query`, and returns the attribute `results` of the `SesameCallback` instance.

■**Note** The Java JAR file `rsesame.jar` also contains the source code for the wrapper class if you want to customize it.

I'll implement the Redland RDF back end in the next section. I prefer using the Sesame back end because of the full RDFS SPARQL query support and because I often deploy to Java environments. Having a Redland back end is still useful, though, especially if you use the C implementation of Ruby.

Designing and Implementing the Ruby and Redland Back End

As you read in Chapter 5, Redland is implemented in the C programming language and the Redland Ruby bindings work with Ruby 1.8.6. The implementation of the class `RedlandBackend` is similar to that of the class `SesameBackend` that you saw in the preceding section. That is, you check whether the environment variable `RDF_DATA` has been set to override the default RDF data directory and call the Redland Ruby bindings for loading RDF data files and performing SPARQL queries:

```
require 'rdf/redland'

class RedlandBackend
  def RedlandBackend.loadRepositories
    if ENV['RDF_DATA']
      rdf_path = ENV['RDF_DATA']
    else
      rdf_path = 'rdf_data'
    end
```

Here, I am using Berkeley DB to store RDF data; see Chapter 5 for other options. The `new='yes'` option deletes any previous Berkeley DB data store and starts with a fresh disk-based hash:

```
    storage=Redland::TripleStore.new("hashes", "test",
                              "new='yes',hash-type='bdb',dir='temp_data'")
    raise "Failed to create RDF storage" if !storage
    @@model=Redland::Model.new(storage)
    raise "Failed to create RDF model" if !@@model
    Dir.entries(rdf_path).each {|fname|
      if fname.index('nt')
        puts "* loading RDF repository: #{rdf_path + '/' + fname}"
        uri=Redland::Uri.new('file:' + rdf_path + '/' + fname)
        parser=Redland::Parser.new("ntriples", "", nil)
        raise "Failed to create RDF parser" if !parser
        stream=parser.parse_as_stream(uri, uri)
        while !stream.end?()
          statement=stream.current()
          @@model.add_statement(statement)
          stream.next()
        end
      end
    }
  end
```

The code to load RDF N-Triple data is similar to what you used in Chapter 5, as is the code using the Redland SPARQL query APIs:

```
  def query sparql_query
    ret = []
    q = Redland::Query.new(sparql_query)
    results=q.execute(@@model)
    while !results.finished?
      temp = []
      for k in 0..results.bindings_count -1
        s = results.binding_value(k).to_s.gsub('[','<').gsub(']','>')
        temp << s
      end
      ret << temp
      results.next
    end
    ret
  end
end
```

■Warning The Redland Ruby bindings support only Ruby 1.8.6, not the newer version Ruby 1.9. The Redland Ruby bindings are not compatible with JRuby. Redland currently does not work with Ruby 1.9.x.

Because the SPARQL endpoint web service accepts simple REST-style queries, it takes only a few lines of Ruby code to implement a client program. I provide an example in the next section.

Implementing a Ruby Test Client

A Ruby client for the SPARQL endpoint portal must URL-encode a SPARQL query, perform an HTTP GET request, and parse JSON-formatted results. You need to have either the native JSON gem or the pure Ruby JSON gem installed to use the following example:

```ruby
require 'open-uri'
require 'cgi'
require 'json'
require 'pp'

server_uri = "http://localhost:4567/?query="
example_query =
    "SELECT ?s ?o WHERE { ?s <http://knowledgebooks.com/ontology/#summary> ?o }"
puts "Example query:\n#{example_query}\n"

loop do
  puts "\nEnter a SPARQL query:"
  line = gets.strip
  break if line.length == 0
  data = open(server_uri + CGI.escape(line)).read
  data = JSON.parse(data)
  puts "Results:"
  pp data
end
```

I print out an example query in this code snippet so you can copy and paste it to get started. You can find this code snippet in the file src/part2/sparql_endpoint_web_service/test_client.rb. In that same directory, you'll find another example client called test_client_benchmark.rb, which I used during development to benchmark and "stress-test" the portal with many simultaneous requests.

Modifying the Portal to Accept New RDF Data in Real Time

The example RDF portal is useful for publishing static data. However, there are times when you'll need to add new data without restarting the portal service. In this section, I will make a few additions to the portal to periodically check whether new files have been added to the rdf_data directory. These modifications will enable the portal to load any new files, after which it can query against the new RDF triples.

If you're using Redland with the Berkeley DB back end or if you're using Sesame, you can then load new files while processing query requests because both Berkeley DB and Sesame provide read/write data-access concurrency support. That said, I am adding my own locking mechanism to pause query processing while new RDF triples are added to the back-end data store.

The SesameBackend and RedlandBackend classes have static methods for loading all RDF N-Triple files from the rdf_data directory to the repository. I need to monkey-patch these classes, adding methods to load individual files.

Monkey-Patching the SesameBackend Class

You need to add the following static method to the class SesameBackend:

```
def SesameBackend.load file_path
  begin
    puts "* loading RDF repository file into Sesame: #{file_path}"
    @@tsm.load_ntriples(file_path)
  rescue
    puts "Error: #{$!}"
  end
end
```

The value of the class attribute @@tsm was set to an instance of SesameWrapper when the file semantic_backend.rb was loaded—that is, when sesame_backend.rb was loaded and the static method SesameBackend.loadRepositories was called.

Loading small RDF N-Triple files is a fast operation, so it won't significantly delay handling REST web-service calls and SPARQL queries.

Monkey-Patching the RedlandBackend Class

You need to add the following static method to the class RedlandBackend:

```
def RedlandBackend.load_new_rdf_file file_path
  begin
    if fname.index('nt')
      puts "* loading RDF file into repository: #{file_path}"
      uri=Redland::Uri.new('file:' + file_path)
      parser=Redland::Parser.new("ntriples", "", nil)
      raise "Failed to create RDF parser" if !parser
      stream=parser.parse_as_stream(uri, uri)
      while !stream.end?
        statement=stream.current
        @@model.add_statement(statement)
        stream.next
      end
    end
  rescue
    puts "Error: #{$!}"
  end
end
```

This is the same code fragment used to load all files in the rdf_data directory when the web-portal service is started. You can simplify the static method RedlandBackend. loadRepositories to use this new method, but I did not make this change in the redland_backend.rb example file.

Modifying the portal.rb Script to Automatically Load New RDF Files

I copied the source file for the script portal.rb to portal_auto_load_files.rb and made three changes. The first change is to define a global lock variable: $lock = false. The second change is to modify the Sinatra get code block to check the lock, and the third change is to add a new thread with a loop at the bottom of the file that periodically looks for and processes new files. The Sinatra framework (with Rack) starts and runs in its own thread.

I added the following checks on the lock variable to the Sinatra get code block:

```
while $lock
  puts "RDF repository locked by writing thread, waiting..."
  sleep(0.1)
end
$lock = true
ses = SemanticBackend.new
response = ses.query(query)
$lock = false
```

For the third change, which consists of additional code to check for new files and process them, I use a hash table stored in the global variable $already_loaded to keep track of RDF N-Triple files that are already in the repository. You do not want to reload files that were already loaded when the portal service started, so you need to add the existing files to this hash table:

```
$already_loaded = {}
Dir.glob('rdf_data/*.nt').each {|file| $already_loaded['rdf_data/' + file] = true}
```

You need to monitor the rdf_data directory, looking for newly added files:

```
work_thread = Thread.new {
  sleep(10) # wait 10 seconds for web portal to start
  file_path = ''
  loop do
    begin
      Dir.glob('rdf_data/*.nt').each {|file_path|
```

If this file is not already loaded, set the lock (waiting, if necessary, for any current queries to be processed) and load the RDF file into the repository:

```
        if !$already_loaded[file_path]
          while $lock
            puts "RDF repository locked by reading thread, waiting..."
            sleep(0.1)
          end
```

```
          $lock = true
          # load file:
          begin
             sleep(1) # wait one second to make sure that the file is fully written
                      # and closed
             SemanticBackend.load_new_rdf_file(file_path)
            puts " * file #{file_path} has been loaded..."
          rescue
               puts "Error calling SemanticBackend.load_new_rdf_file(➥
  #{file_path}): #{$!}"
            end
          $lock = false
          $already_loaded[file_path] = true
        end
      }
    rescue
      puts "Error adding new file #{file_path} to RDF Reository: #{$!}"
    end
    sleep(10) # wait for 10 seconds before checking for new RDF files
  end
}
```

In this code example, I wait one second after noticing a new file before reading it. For very large files, you might want to change this delay to two seconds to be sure that the new file is closed. I tested this version of the portal by running three simultaneous instances of the test_client_benchmark.rb script using both Ruby 1.8.6 and JRuby.

Modifying the Portal to Generate Graphviz RDF Diagrams

Whereas RDF data is designed for use by software, you have seen that formatting RDF in N3 format makes it much easier for people to read and understand. But when you read through an N3 file, it isn't easy to recognize common subject or object URIs. This is when you can benefit from showing RDF data as graphs. Graphs can help you visualize common triple subjects, for example, because common subjects and objects are represented by a single node in a graph. The Graphviz program (available at http://www.graphviz.org/) is an effective tool you can use to lay out RDF graphs for human viewing, so you'll use it again in this section.

I use the ruby-graphviz gem that you can install in Ruby 1.8.6 like this:

```
gem install ruby-graphviz
```

The Graphviz input dot file format is simple, and when I use Graphviz in Common Lisp visualization applications (which I do often), I directly write out dot files. On a few occasions I have done this in Ruby also. That said, using the high-level ruby-graphviz gem library is simpler, so you'll use it in this example. The ruby-graphviz gem is not currently compatible with

JRuby, so writing dot files for input to the Graphviz application is a good option if your application uses JRuby.

This example supports two output formats: PNG graphic files and Graphviz dot files. I also support full URI display and a concise node display. This example supports a few output options, which you set by defining "concise" and/or "dot" in the REST query string:

- "http://localhost:4567/?query=" for full node names and the default PNG file type

- "http://localhost:4567/?dot&query=" for full node names with the Graphviz dot file type

- "http://localhost:4567/?concise&query=" for concise node names and the default PNG file type

- "http://localhost:4567/?concise&dot&query=" for concise node names with the Graphviz dot file type

Figures 7-3 and 7-4 contain graph visualizations of RDF query results: the former shows an example with full node names, and the latter shows the same example with concise node names. In each graph visualization, the nodes on the left side represent RDF triple subjects—URLs of news stories, in this case.

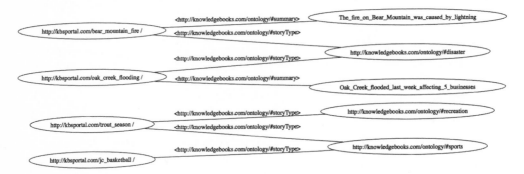

Figure 7-3. *Graphviz visualization of sample data in the rdf_data directory*

A simple SPARQL query to return all triples in the test data store produced Figures 7-3 and 7-4:

```
SELECT ?subject ?predicate ?object { ?subject ?predicate ?object }
```

You can use any SPARQL query that returns three values in each result.

I created the Ruby source file portal_graphviz.rb by cloning the portal.rb script, adding support for using query results to generate calls to the ruby-graphviz library, and then returning either PNG image data or Graphviz dot text data as the results of web-service calls. This new script is long, so I will intersperse explanations with the code snippets. This script is released under the GPL because the graphviz gem is released under the GPL:

```
require 'rubygems'
require 'sinatra'
require 'json'
```

```
require 'cgi'
require 'ftools'
require 'lib/sparql_endpoint_web_service'
require 'graphviz'
# License: GPL version 2 or later (because graphviz gem is GPL)
require 'pp'
```

The Sinatra get utility method gets the query string out of the HTTP request and looks for the flags that specify whether to use a concise node display format and whether to use the default PNG format or return Graphviz dot text data:

```
get '/*' do
```

I ignore any returned results that do not contain three items:

```
  if params['query']
    query = CGI.unescape(params['query'])
    concise = params.key?('concise')
```

You use the Sinatra method content_type to set the HTTP response header for the appropriate Multipurpose Internet Mail Extensions (MIME) type:

```
    if params.key?('dot')
      content_type 'text/dot'
      file_type = 'dot'
    else
      content_type 'image/png'
      file_type = 'png'
    end
    ses = SemanticBackend.new
    response = ses.query(query)
```

The response from a SPARQL query is an array containing subarrays of strings. These subarrays need to have three elements (representing a triple), or they are ignored when the RDF graph visualizations are being generated. I use a new utility function called write_graph_file to generate an output file that is then read and returned to the requesting client. The Ruby graphviz library does not support writing to memory, so I use a temporary file that I delete as soon as I am done using it:

```
    write_graph_file(response, concise, file_type)
    data = File.open("temp_data/gviz_#{$counter}.#{file_type}").read
    File.delete("temp_data/gviz_#{$counter}.#{file_type}")
    data
  end
end
```

I need unique temporary file names (I cannot use the standard Ruby TempFile class because the graphviz gem requires a file path as a string and not a file object), so I generate a small random number to add to the generated file name:

```
$counter = rand(10000) # gets reset for every request
```

When you run Sinatra in development mode, the source code for the script is reread for every web-service request. Because this counter is reinitialized for each new web-service request, I use a random number for a counter to distinguish temporary output files. When running Sinatra in production mode, you can just use a counter if you wish.

The following utility function's arguments are a response (expected to be an array of triples) and optional flags for concise node names and PNG or Graphviz output formats. You use the ruby-graphviz gem library to create an in-memory graph and then write the graph to an output file:

```
def write_graph_file response, concise_names=false, file_type='png'
  g = GraphViz::new( "G", "output" => file_type )
  g["rankdir"] = "LR"
  g.node["shape"] = "ellipse"
  g.edge["arrowhead"] = "normal"
  response.each {|triple|
```

I ignore any results that do not contain three strings representing either URIs or literals:

```
    if triple.length == 3
      if concise_names
        3.times {|i| triple[i] = concise_name_for(triple[i]); puts triple[i]}
      end
```

The third result value might be a string containing spaces. Space characters signal Graphviz to create new nodes, so I replace spaces with the underscore character (you see the effect of this in Figures 7-3 and 7-4):

```
        triple[2] = triple[2].gsub(' ', '_')
        g.add_node(triple[0])
        g.add_node(triple[2])
        g.add_edge( triple[0], triple[2], "arrowhead" => triple[2], ➥
:label => triple[1] )
      end
  }
  fpath = "temp_data/gviz_#{$counter}.#{file_type}"
  g.output( :file => fpath )
  puts "Output is in:   #{fpath}  counter: #{$counter}"
end
```

The utility function concise_name_for converts full URIs to shorter (concise) names:

```
def concise_name_for str
  if str[0..4] == "<http" && (index = str.gsub('/>',' ').rindex("/"))
    str[index+1..-2].gsub('#','').gsub('/',' ').strip
  else
    str
  end
end
```

The concise_name_for function checks whether the input string looks like a full URI, and if so, strips off the characters after the last "/" character. Figure 7-4 shows the RDF visualization graph using concise names produced by the function concise_name_for.

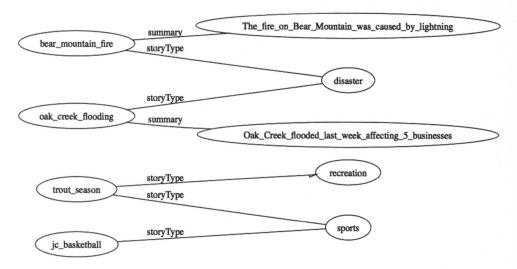

Figure 7-4. *Graphviz visualization with concise node names of sample data in the rdf_data directory*

The Graphviz format is convenient because you can use the Graphviz utility programs for viewing and converting to many different graphic formats. Here is a snippet from the generated Graphviz file for Figure 7-4. It defines some graph parameters, two nodes, and one link between the two nodes:

```
digraph G {
    graph [rankdir=LR];
    node [label="\N", shape=ellipse];
    trout_season [];
    recreation [];
    trout_season -> recreation [arrowhead=recreation, label=storyType];
}
```

Warning The Ruby graphviz gem is not currently available to work with JRuby because it uses native extensions, so run this example using Ruby 1.8.6. If you need to use JRuby, you can directly write out dot input files for Graphviz—I leave this as an exercise for the reader.

Using RDF graph visualization is especially useful when you are explaining RDF and Semantic Web technologies to people. In addition to "factoring out" common RDF subjects and objects as the same graph node, the graph also makes it easier to understand the pattern matching used in processing SPARQL queries.

Wrapup

REST-style web services generally provide a convenient way to combine systems implemented in different programming languages and running on different operating system platforms. Web services also support systems that are distributed over geographically dispersed servers. I'll continue to use web services in future examples. The web-service examples in this chapter serve as examples for publishing RDF data.

Part 2 introduced you to the core Semantic Web technologies for developing Web 3.0 applications. This is a large topic, and I concentrated on practical issues to get you started writing applications. Whereas knowledge of RDF and RDFS is sufficient for many applications, you might eventually want to study OWL modeling and more advanced inferencing techniques.

Part 3 of this book covers techniques for information gathering and storage. You'll continue using the Semantic Web technologies covered in Part 2, and you'll also learn of a good alternative to the custom SPARQL endpoint developed in this chapter: the D2R system. D2R, which I'll cover in Chapter 11, provides a SPARQL interface for existing relational databases.

PART 3

■ ■ ■

Information Gathering and Storage

I covered Semantic Web technologies in Part 2. You'll continue using these technologies, but in Part 3 you'll also use relational databases to perform free-text search and to store data scraped from web sites. Plus, you'll learn to use a wrapper to publish relational data as RDF with a SPARQL interface, to use and produce Linked Data resources, and to implement strategies for large-scale data storage.

CHAPTER 8

■■■

Working with Relational Databases

Even if you do not develop Web 3.0 applications, the material in this chapter will save you time and effort when working with relational databases. Specifically, I'll discuss object-relational mapping (ORM) and provide a tutorial on two effective ORM frameworks: ActiveRecord and DataMapper. I think that a trend in Web 3.0 development will be a continued reduction in the cost of development and software maintenance. I expect to see continued improvements in the expressiveness of ORM frameworks based on domain-specific languages (DSLs), which will make development more efficient.

It is worth explaining my own preferences regarding whether to use an ORM tool or whether to use direct SQL queries when working with relational databases in Ruby applications. Largely because Ruby is a dynamic language, it was possible for the developers of ActiveRecord and DataMapper to create DSLs that drastically reduce the amount of code required to use relational databases. I think the fact that Ruby is a dynamic language that is also well-suited to creating DSLs is the reason Ruby ORM tools are so effective. Java is not a dynamic language, on the other hand, and for this reason the developers of Hibernate—the best Java ORM tool—could not create a DSL-like solution for ORM. So, whereas I almost always use ActiveRecord or DataMapper when doing Ruby development, I rarely bother using Hibernate when doing Java development.

Note For the purposes of this chapter, I'm assuming that you are a Ruby developer but not necessarily a Ruby on Rails developer. If you have not experienced Rails development, see Chapter 15 for two complete examples of using Ruby on Rails.

ActiveRecord and DataMapper take different approaches to domain modeling. ActiveRecord uses the metadata in a database to automatically generate Ruby model classes, while with DataMapper you design your Ruby classes and have DataMapper create database schemas for you. You have probably already used ActiveRecord because it's the default ORM tool for Ruby on Rails, but you might not have tried the less widely used DataMapper yet. DataMapper is often used with the Merb web-application framework. Because Merb is getting merged into Rails 3.0, DataMapper might become more popular with Rails developers.

Figure 8-1 shows the basic concept of ORM: database rows are mapped to objects, and database columns are mapped to object attributes.

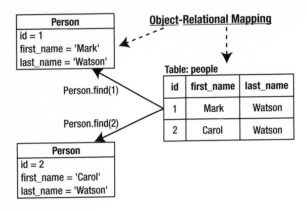

Figure 8-1. *Object-relational mapping*

PostgreSQL and MySQL are my "workhorse" database systems. I seldom use SQLite except in standalone single-process and single-threaded applications. That said, SQLite is fine for simple schemas and offers the advantage of easy use and installation for demos and other quick projects. ActiveRecord and DataMapper mostly abstract away differences between SQLite, PostgreSQL, and MySQL. (I will not cover SQLite in this chapter.)

While you might have used ActiveRecord in writing Ruby on Rails applications and DataMapper in writing Merb web applications, you might not have used ActiveRecord or DataMapper by itself (that is, separate from a Rails or Merb project). I'll exclusively use both ActiveRecord and DataMapper as standalone entities in this chapter. I will use ActiveRecord in two Rails applications in Chapter 15.

While I'll concentrate on relational-database storage solutions in this chapter, in Chapter 12 I'll demonstrate alternative scalable storage solutions such as Amazon Simple Storage Service (Amazon S3) in conjunction with Amazon Elastic Compute Cloud (EC2), and memcached.

Doing ORM with ActiveRecord

Because ActiveRecord is the default ORM tool used in Ruby on Rails, you benefit from many contributors to its code base and a great deal of documentation. Whenever I choose languages and frameworks for projects, I simply try to use the best tool for each job. However, I also favor tools that are widely used so customers have more flexibility in finding developers to work on and maintain their projects. While both ActiveRecord and DataMapper are excellent tools, I usually favor ActiveRecord because it is more commonly used.

The following is an introduction to using ActiveRecord in standalone, or embedded, mode. My purpose here is not to provide a complete reference, but to show you the techniques that I use most often.

Quick-Start Tutorial

I usually use Ruby on Rails migrations to specify the attributes for my model classes and let these Rails migrations automatically generate the database schema and create tables corresponding to each model class. If you are a Rails developer, you should already be familiar with this process. But for the purposes of this chapter, I'll use ActiveRecord to read the metadata for existing databases and generate model classes automatically. I will use database migrations in Chapter 15.

I am going to use a common setup for the examples in this section; you can find this setup code in the file src/part3/ar_setup_example.rb (download the code samples for this chapter from the Source Code/Download area of the Apress web site at http://www.apress.com). The static class method ActiveRecord::Schema.define supports a similar syntax for describing model classes and their attributes, as seen in the ar_setup_example.rb script:

```
require 'activerecord'

ActiveRecord::Base.establish_connection(:adapter  => :mysql, :database => "test")

ActiveRecord::Schema.define do
  create_table :news_articles do |t|
    t.string :url
    t.string :title
    t.string :summary
    t.string :contents
  end

  create_table :people do |t|
    t.string :name
    t.integer :news_article_id
  end

  create_table 'places' do |t|
    t.string 'name'
    t.integer :news_article_id
  end
end
```

This script creates three database tables: news_articles, people, and places. Using ActiveRecord::Schema is very similar to using Rails migrations, which I will do in Chapter 15. The file ar_classes_from_database_metadata.rb contains the following example of creating model classes NewsArticle, Person, and Place from the three database tables that we just created:

```
require 'activerecord'

ActiveRecord::Base.establish_connection(:adapter  => :mysql, :database => "test")
```

I have now opened a connection to the database test that contains the tables news_articles, people, and places. I'll continue with this example in the next section, where you will see how to handle one-to-many relationships.

One-to-Many Relationships

ActiveRecord has built-in support for handling associations between classes. An example would be a class containing one or more instances of another class. I will define a class called NewsArticle from the metadata for the news_articles table and specify that an instance of this class can contain zero or more people and places:

```
class NewsArticle < ActiveRecord::Base
  has_many :people
  has_many :places
end
```

Now I'll specify that instances of the classes Person and Place can belong to a news article:

```
class Person < ActiveRecord::Base
  belongs_to :news_article
end

class Place < ActiveRecord::Base
  belongs_to :news_article
end
```

After creating an instance of the class NewsArticle and saving it to the database, I can query the class to return all articles, query the class to return a single article with a specific primary key id, or use a query condition:

```
NewsArticle.new(:url => 'http://test.com/bigwave',
                :title => 'Tidal Wave Misses Hawaii',
                :summary => 'Tidal wave missed Hawaii by 500 miles',
                :contents => 'A large tidal wave travelled across
                              the pacific, missing
                              Hawaii by 500 miles').save
```

This new object would not be persisted to the news_articles table until the save method is called. Here I fetch all news articles and then fetch just the news article with the primary key id equal to 1:

```
all_articles = NewsArticle.find(:all) # or: NewsArticle.all
wave_article = NewsArticle.find(1)    # or: NewsArticle.first
```

I can also fetch the same article, overwriting the previous value in the variable wave_article by using a conditional match on the title column:

```
title = 'Tidal Wave Misses Hawaii'
wave_article = NewsArticle.find(:all, :conditions => ['title = ?', title])
```

This news-article object prints as:

```
#<NewsArticle id: 1, url: "http://test.com/bigwave", title: "Tidal Wave Misses ➥
Hawaii", summary: "Tidal wave missed Hawaii by 500 miles", contents: ➥
"A large tidal wave travelled across the pacific, missing Hawaii by 500 miles">
```

Currently this object is not associated with instances of classes Person and Place:

```
irb(main):056:0> wave_article.people
=> []
irb(main):011:0> wave_article.places
=> []
```

You can use the following code snippets to create a person object and a place object that you can add to this news story later:

```
mark = Person.new(:name => "Mark")
mark.save

sedona = Place.new(:name => "Sedona Arizona")
sedona.save
```

The following Ruby statement adds the place sedona to the places collection in the news article and also sets the attribute news_article_id: of the place object to the id of the news article:

```
wave_article.places << sedona
```

You can also add a person object and save the article to the database:

```
wave_article.people << mark
wave_article.save
```

I actually did more work than I needed to in this example; there is no need to save the person and place objects because the news-article object was already saved. In other words, adding person and place objects to the news story also saves the person and place objects to the database. Here's the preceding example again, in a shorter form:

```
NewsArticle.new(:url => 'http://test.com/bigwave',
                    :title => 'Tidal Wave Misses Hawaii',
                    :summary => 'Tidal wave missed Hawaii by 500 miles',
                    :contents => 'A large tidal wave travelled across
                                        the pacific, missing
                                        Hawaii by 500 miles').save
wave_article = NewsArticle.find(1)
wave_article.places << Place.new(:name => "Sedona Arizona")
wave_article.people << Person.new(:name => "Mark")
```

ActiveRecord defines a lot of functionality for classes generated from database metadata. For the class NewsArticle, it defines 140 public methods, the most useful of which are these:

---

after_create, after_destroy, attribute_names, attributes, before_create, ➥
before_destroy, before_save, before_update, changes, **contents, contents=,** ➥
contents?, delete, from_json, from_xml, **id, id=,** people, person_ids, places, ➥
place_ids, reload, save, save!, **summary, summary=, summary?, title, title=,** ➥
title?, to_json, to_xml, transaction, **url, url=, url?**

I highlighted in bold font methods for the class attributes contents, id, summary, title, and url. The attribute methods ending with a question mark return a true value if the attribute is defined for an object. I underlined the two methods representing the collections of associated person and place objects. If you are developing web services, then you will likely use the methods for encoding to JSON and XML, and decoding back to Ruby objects.

In the conditional find method call NewsArticle.find(:all, :conditions => ['title = ?', title]), I matched the title with an exact string match. ActiveRecord, in good DSL form, provides a nice notation for performing SQL "like" matching for databases supporting "like" queries. For example, to match any news article with the word "wave" (equivalent to a '%wave%' match), I could use the condition:

```
NewsArticle.find(:all, :conditions => ["title like ?", '%wave%'])
```

You'll see later in this chapter how to use the event-callback methods before_create, after_create, before_destroy, and after_destroy (see the section, "Handling Callbacks and Observers in ActiveRecord").

Handling Transactions

ActiveRecord associates transactions with a database connection, so it is possible to modify instances of different Ruby ActiveRecord model classes inside a single transaction. You can find this section's examples in the Ruby script file ar_transactions_demo.rb, and I'll also show the file's contents here, interspersed with explanations:

```
require 'activerecord'

ActiveRecord::Base.establish_connection(:adapter  => :mysql, :database => "test")

class NewsArticle < ActiveRecord::Base
  has_many :people
  has_many :places
end
class Person < ActiveRecord::Base
  belongs_to :news_article
end
class Place < ActiveRecord::Base
  belongs_to :news_article
end

require 'pp'
```

I am assuming that you've set up the database using this chapter's previous examples, so that you can create these three objects by using the first row in each of the tables news_articles, people, and places:

```
mark = Person.find(1)
sedona = Place.find(1)

pp mark
pp sedona
```

The output from these three "pretty-print" statements is:

```
#<Person id: 1, name: "Mark", news_article_id: 1>
#<Place id: 1, name: "Sedona Arizona", news_article_id: 1>
```

I will change the names of the objects in memory, but this has no effect on the data in the database until either of these objects is saved, or until an object containing these objects is saved:

```
mark.name = "Mark Watson"
sedona.name = "Sedona, Arizona"
```

You can use transaction blocks with the static class method `ActiveRecord::Base.transaction`. Executing both of the following save operations inside a transaction block ensures that either both save operations affect the database, or that neither of them affects the database:

```
ActiveRecord::Base.transaction do
  mark.save!
  sedona.save!
end
```

You can also define a transaction block using the instance method `transaction` of any model-class instance:

```
mark.transaction do
  mark.save!
  # perform some other operation that might fail here...
end
```

Save and delete actions are automatically wrapped in a transaction block. This means that any event-handling callbacks or observer code is run under the protection of a transaction. You can define callbacks to be run both before and after a model object is saved to a database. If an error occurs while you're running callback or observer code, then no effects on the database occur. I'll cover callbacks and observers next.

Handling Callbacks and Observers in ActiveRecord

Callbacks and observers serve similar functions, but you'll choose which one to implement depending on the situation. Callbacks are implemented as part of a model class, while observers are implemented as separate classes from the models that they monitor.

ActiveRecord supports 12 types of callback hooks. The most common hooks are used before and after creating a model object, and before and after saving a model object. I will modify the model class `Place` from our examples to implement these four callbacks. The following example snippets are in the file `ar_callback_demo.rb`:

```
require 'activerecord'

ActiveRecord::Base.establish_connection(:adapter  => :mysql, :database => "test")

class Place < ActiveRecord::Base
```

The ActiveRecord base class includes the following methods, which I use to register methods that I define as callback functions:

```
before_create :monitor_before_place_creation
after_create :monitor_after_place_creation
before_save :monitor_before_place_save
after_save :monitor_after_place_save
before_destroy :monitor_before_place_destroy
after_destroy :monitor_after_place_destroy
```

The ActiveRecord runtime system will call any methods that you register at the appropriate times.

Because callback methods do not get called with any arguments, they are limited to working with internal attributes and attributes of other objects that they have access to. The following callback methods print their names so that you know when they are called (not a very practical example, but useful for understanding when they are called):

```
private
  def monitor_before_place_creation; puts 'monitor_before_place_creation'; end
  def monitor_after_place_creation;  puts 'monitor_after_place_creation';  end
  def monitor_before_place_save;     puts 'monitor_before_place_save';     end
  def monitor_after_place_save;      puts 'monitor_after_place_save';      end
  def monitor_before_place_destroy;  puts 'monitor_before_place_destroy';  end
  def monitor_after_place_destroy;   puts 'monitor_after_place_destroy';   end

end

puts "Create a new in-memory place:"
place = Place.new(:name => 'Arizona')
puts "Save the new in-memory place to the database:"
place.save!
puts "Destroy the object and remove from database:"
place.destroy
```

The output is:

```
$ ruby ar_callback_demo.rb
Create a new in-memory place:
Save the new in-memory place to the database:
monitor_before_place_save
monitor_before_place_creation
monitor_after_place_creation
monitor_after_place_save
Destroy the object and remove from database:
monitor_before_place_destroy
monitor_after_place_destroy
```

Observers offer the advantage of not requiring changes to model classes. You should make the decision to use callbacks or observers by asking yourself if the monitoring actions have anything to do with behavior that relates to the object itself, or if the monitoring actions have nothing to do with the functionality that you would associate with a model class.

I suggest using callbacks when the desired monitoring functionality involves actions like verifying attribute values before saving an object, compressing data before storing an object in a database, or calculating undefined attribute values before a save operation.

I recommend using observers when the monitoring functionality has little to do with an object's internal state or behavior, or when a description of a model class and its function differs from a description of the functionality of the monitoring process. Examples for using observers include logging object events to a log file and saving an object state to a remote data store whenever a local database save operation is performed. The use of an observer allows you to change monitoring behavior without modifying a model class. This is a good separation of concerns. The file ar_observer_demo.rb contains the following code snippets:

```
require 'activerecord'

ActiveRecord::Base.establish_connection(:adapter  => :mysql, :database => "test")

class Place < ActiveRecord::Base
  belongs_to :news_article
end
class Person < ActiveRecord::Base
  belongs_to :news_article
end
```

If I were to name a subclass of ActiveRecord::Observer PlaceObserver, then I would not have to explicitly state the class or classes that my observer class is observing. However, I want the following class to observe two classes (Place and Person), so I need to explicitly use the observe method:

```
class MyObserver < ActiveRecord::Observer
  observe :place, :person
  def before_save(model)
    puts "** Before saving #{model}"
  end
  def after_save(model)
    puts "** After saving #{model}"
  end
end
```

You can define methods in your observer classes for the same events as callback hooks. I just defined before_save and after_save in MyObserver, but you could also define before_destroy, after_destroy, before_create, after_create, and so on.

If you were developing a Ruby on Rails application, then you'd use a configuration setting to add observer classes to the ActiveRecord::Base.observers list, after which Rails would instantiate the observers during application initialization. However, since you're using ActiveRecord by itself, without Rails, you need the following two lines of code to register an observer class and instantiate it:

```
ActiveRecord::Base.observers << MyObserver
ActiveRecord::Base.instantiate_observers
```

```
puts "Create a new in-memory place:"
place = Place.new(:name => 'Arizona')
puts "Save the new in-memory place to the database:"
place.save!
puts "Destroy the place object and remove from database:"
place.destroy

puts "Create a new in-memory person:"
brady = Person.new(:name => 'Brady')
puts "Save the new in-memory person to the database:"
brady.save!
puts "Destroy the person object and remove from database:"
brady.destroy
```

The output from this script is:

```
$ ruby ar_observer_demo.rb
Create a new in-memory place:
Save the new in-memory place to the database:
** Before saving #<Place:0x136dd90>
** After saving #<Place:0x136dd90>
Destroy the place object and remove from database:
Create a new in-memory person:
Save the new in-memory person to the database:
** Before saving #<Person:0x7fdcbc>
** After saving #<Person:0x7fdcbc>
Destroy the person object and remove from database:
```

Using ActiveRecord callbacks and observers allows you to monitor object and database events without having to modify ActiveRecord itself, subclass it, or make monkey patches.

Modifying Default Behavior

ActiveRecord makes a lot of assumptions that usually make your development easier. However, there are a few common situations when you'd want to override default behavior.

If you need to use ActiveRecord with an existing database that does not follow the assumption of a primary key named id and/or a table name that is the plural form of a model-class name, you can easily override these defaults. For example, if you have an existing database table named people_in_arizona with a primary key of person_id, then you could set up a model class to use this table, overriding the default mapping from class name to table name:

```
class Person < ActiveRecord::Base
  set_table_name 'people_in_arizona'
  set_primary_key 'person_id'
end
```

■Tip Peter Szinek recommends Dr. Nic Williams's Magic Model Generator (`http://magicmodels.`
`rubyforge.org/magic_model_generator/`) for dealing with large legacy databases with many intercon-
nected tables. (Currently it works only with Ruby 1.8.6.) A time-saver!

Another default that you will sometimes want to override for efficiency reasons is Active-
Record's lazy loading. For example, suppose you're using the `NewsArticle`, `Place`, and `Person`
classes once again; recall that an instance of `NewsArticle` can contain zero or more people
and places. However, by default, the associated people and places are not immediately loaded
when you create a news-article model object. The following code snippets reside in the file
`ar_lazy_loading_demo.rb`:

```
require 'activerecord'
```

```
ActiveRecord::Base.establish_connection(:adapter  => :mysql, :database => "test")
```

You want to see all SQL queries that ActiveRecord makes to the test database, so enable
logging of all events to a logger that writes to standard output:

```
ActiveRecord::Base.logger = Logger.new(STDOUT)

class NewsArticle < ActiveRecord::Base
  has_many :people
  has_many :places
end

class Place < ActiveRecord::Base
  belongs_to :news_article
end
class Person < ActiveRecord::Base
  belongs_to :news_article
end

puts "Fetch news article from row 1 of database:"
news = NewsArticle.find(1)
puts "Access all people in news article:"
people = news.people
puts "Access all places in news article:"
places = news.places
```

The output is:

```
$ ruby ar_lazy_loading_demo.rb
Fetch news article from row 1 of database:
  SQL (0.3ms)   SET SQL_AUTO_IS_NULL=0
  NewsArticle Columns (1.9ms)   SHOW FIELDS FROM `news_articles`
  NewsArticle Load (1.3ms)   SELECT * FROM `news_articles` WHERE ➡
```

```
(`news_articles`.`id` = 1)
Access all people in news article:
 Person Load (0.4ms)   SELECT * FROM `people` WHERE (`people`.news_article_id = 1)
Access all places in news article:
  Person Columns (1.7ms)   SHOW FIELDS FROM `people`
[#<Person id: 1, name: "Mark Watson", news_article_id: 1>]
  Place Load (0.3ms)   SELECT * FROM `places` WHERE (`places`.news_article_id = 1)
  Place Columns (1.6ms)   SHOW FIELDS FROM `places`
```

Now rerun the script, but turn off lazy loading for people and places:

```
puts "Load a news article with lazy loading turned off:"
news2 = NewsArticle.find(1, :include => [:people, :places])
```

Here's the output with lazy loading turned off:

```
$ ruby ar_lazy_loading_demo.rb
Load a news article with lazy loading turned off:
  NewsArticle Load (0.3ms)   SELECT * FROM `news_articles` WHERE ➡
(`news_articles`.`id` = 1)
  Person Load (0.7ms)   SELECT `people`.* FROM `people` WHERE ➡
(`people`.news_article_id = 1)
  Person Columns (3.9ms)   SHOW FIELDS FROM `people`
  Place Load (0.4ms)   SELECT `places`.* FROM `places` WHERE ➡
(`places`.news_article_id = 1)
  Place Columns (1.5ms)   SHOW FIELDS FROM `places`
```

You can see that with lazy loading turned off, ActiveRecord makes all queries for associated objects at once.

Using SQL Queries

ActiveRecord supports using native SQL queries, but there are good reasons not to use them directly. Differences in relational databases are largely abstracted away by the custom connectors for each type of supported database. If you write native SQL queries, your code is much less likely to be portable between different database engines. Your application will also be more difficult to read and understand.

If you need to use native SQL with ActiveRecord, though, you can easily do so. Start by creating a class Place from the database table places and declare that this class belongs to the class NewsArticle (the following code snippets are in the file ar_execute_sql_demo.rb):

```
require 'activerecord'
require 'pp'
ActiveRecord::Base.establish_connection(:adapter  => :mysql, :database => "test")
class Place < ActiveRecord::Base; belongs_to :news_article; end
```

This example SQL query fetches all rows, returning data for every column:

```
sql = "select * from places"
results = ActiveRecord::Base.connection.execute(sql)
```

The value of the variable results is `Mysql::Result` because you're using a MySQL connector. You still have portability between different database systems because a `Results` class is defined for each database connector and they all implement the same access methods. The following statement prints the column (or field) names:

```
pp results.fetch_fields.collect {|f| f.name}
```

The output is:

```
["id", "name", "news_article_id"]
```

The following statement iterates over all rows in the results:

```
results.each {|result| pp result}
```

Each row is provided as an array of string values. Note that using ActiveRecord is much more convenient than using native SQL in this case, because it converts numeric values to Ruby numeric data types (integer or float, as appropriate).

The output of the preceding statement is:

```
["1", "Sedona, Arizona", "1"]
```

If you select a subset of columns in a query, then the returned column names simply match the requested columns:

```
sql = "select name, news_article_id from places"
results = ActiveRecord::Base.connection.execute(sql)
pp results.fetch_fields.collect {|f| f.name}
results.each {|result| pp result}
```

The output is:

```
["name", "news_article_id"]
["Sedona, Arizona", "1"]
```

I almost never use direct SQL with ActiveRecord if it is possible to perform the required queries with the ActiveRecord `find` method. I have also mostly stopped using direct interfaces to PostgreSQL and MySQL because using ActiveRecord and DataMapper is so much more convenient.

Accessing Metadata

I have used Java for enterprise development since Java 1.0 and I have written several Java programming books. I still favor Java for some types of development requiring very high performance, but I believe that comparing access to database metadata using Ruby with ActiveRecord vs. using Java with Hibernate (my favorite Java ORM) makes a strong argument for the superiority of using a dynamic scripting language like Ruby (or Python) for development. Hibernate offers complete and well-designed metadata APIs, but accessing metadata using ActiveRecord is much simpler in comparison. To drive home a point: for reducing software-development effort, it is much more efficient to use a dynamic language like Ruby, with its well-designed and well-implemented higher-level DSLs that are tailored to specific types of applications.

The following code snippets are in the file ar_metadata_demo.rb. Here I'll demonstrate the ActiveRecord methods table_name, column_names, and count:

```
require 'activerecord'
require 'pp'

ActiveRecord::Base.establish_connection(:adapter  => :mysql, :database => "test")
class Place < ActiveRecord::Base;  belongs_to :news_article;  end

pp Place.table_name       # returns the table name for this model class
pp Place.column_names   # returns an array of strings for column names
pp Place.count                # returns the row count of the places table
```

Here is output of the preceding code snippet, which shows the table name for a model class, the database column names, and the number of rows in the table:

```
"places"
["id", "name", "news_article_id"]
1
```

I have emphasized aspects of ActiveRecord most useful for non-Rails applications. ActiveRecord, which was designed for use with Rails, supports validation functionality that is tailored to handling common tasks such as verifying HTML form data. ActiveRecord also has API hooks for optimizing caching. I use ActiveRecord frequently in my work and if you have not yet tried it, I hope that my introduction to it has both "shown you the good stuff" and convinced you to give it a try. The next section introduces you to another good ORM system: DataMapper.

Doing ORM with DataMapper

You'll find a good resource in the DataMapper web site (http://datamapper.org), which contains both code and documentation. In this section, I will introduce you to DataMapper in much the same way I introduced you to ActiveRecord: by showing you the features that I find most useful in my own work.

DataMapper takes a more Ruby-like approach to ORM. One DataMapper feature that differs greatly from ActiveRecord's technique involves *object identity*. If you perform two ActiveRecord find searches and generate two objects from the same table row of data, then these two Ruby objects are different. In the same scenario, DataMapper will produce two references to the same object. This makes DataMapper more efficient, and the one-to-one mapping between rows in a database and objects instantiated in memory might eliminate some errors caused by the confusion of having two different objects representing the same

back-end data. DataMapper is both faster than ActiveRecord and uses less memory. Data-
Mapper also is relatively easy to use as a front end for nonrelational data stores such as
CouchDB, flat files, and so on.

Here's how to install DataMapper with Data Objects back ends for SQLite, PostgreSQL,
and MySQL:

```
gem install do_sqlite3 do_postgres do_mysql dm-core
```

■**Note** DataMapper depends on the Data Objects gems (one each for supported databases). Most of the
Data Objects gems require native C language extensions and will not work with JRuby.

As you will see, DataMapper has some useful features that differentiate it from Active-
Record. I do not often use DataMapper because there is currently no official JRuby support.
Looking forward, I expect that DataMapper will become my Ruby ORM of choice when it is
supported on JRuby.

Quick-Start Tutorial

For comparison with ActiveRecord, I'll use the same three database tables that I used for the
ActiveRecord tutorial. There is a difference, however, in modeling data: DataMapper takes a
Ruby-centric approach, in which you design model classes and annotate them to generate
database tables.

The following code snippets are in the file dm_setup_example.rb:

```
require 'rubygems'  # required for Ruby 1.8.6
require 'dm-core'
require 'pp'
```

I am using MySQL in this example, but you can change the following DataMapper setup
calls:

```
#DataMapper.setup(:default, 'sqlite3::memory:')
#DataMapper.setup(:default, "sqlite3:temp_data/dm_test.db")
DataMapper.setup(:default, 'mysql://localhost/test')
#DataMapper.setup(:default, 'postgres://localhost/test')
```

You start DataMapper model-class definitions by including (mixing in) the DataMapper
Resource module. You define class attributes using the DataMapper function property, which
takes two required arguments (and optional arguments not used here) to specify the attribute
name and the attribute type. As you saw in ActiveRecord, specifying an attribute type enables
automatic conversion of string values to the appropriate Ruby types when you create objects
in memory from rows in a database table. The difference between String and Text types is
that Text attributes are assumed to be large and their values are not immediately fetched from
a database for multiple-result queries (a situation known as lazy loading, which I discussed
earlier):

```
class NewsArticle
  include DataMapper::Resource
  property :id,       Serial
  property :url,      String
  property :title,    String
  property :summary,  String
  property :contents, Text

  has n, :people
  has n, :places
end
```

You use the DataMapper method has to define associations. Here I am stating that a news article can contain zero or more people and places. For example, if I wanted to allow only one associated place, then I could have used has 1, :places. In the following class definition for Person, I specify that a person object can belong to a news article using the statement belongs_ to :news_article:

```
class Person
  include DataMapper::Resource
  property :id,   Serial
  property :name, String

  belongs_to :news_article
end

class Place
  include DataMapper::Resource
  property :id,   Serial
  property :name, String

  belongs_to :news_article
end
```

Note that the containing class uses a plural form of contained class names, while the contained class uses a singular form for referring to the containing class.

You can change the following code to modify the level of detail for logged information:

```
#DataMapper::Logger.new(STDOUT, :off)
#DataMapper::Logger.new(STDOUT, :fatal)
#DataMapper::Logger.new(STDOUT, :error)
#DataMapper::Logger.new(STDOUT, :warn)
#DataMapper::Logger.new(STDOUT, :info)
DataMapper::Logger.new(STDOUT, :debug)
```

The following statement creates all tables for the DataMapper model classes that we have defined (this deletes all existing data!):

```
DataMapper.auto_migrate!
```

DataMapper model classes provide a new method for instance creation. You can create an empty object (that is, an object with no attribute values set) or, as I do here, specify attribute values using a hash:

```
news1 = NewsArticle.new(:url => 'http://test.com/bigwave',
                        :title => 'Tidal Wave Misses Hawaii',
                        :summary => 'Tidal wave missed Hawaii by 500 miles',
                        :contents => 'A large tidal wave travelled across
                                         the pacific,
                                         missing Hawaii by 500 miles')
news1.save
```

The following code snippets demonstrate lazy loading of Text properties:

```
news_articles =  NewsArticle.all
pp news_articles
```

The output does not show the contents attribute because properties of type Text are lazily loaded:

```
[#<NewsArticle id=1 url="http://test.com/bigwave" title="➥
Tidal Wave Misses Hawaii" summary="Tidal wave missed Hawaii by 500 miles" ➥
contents=<not loaded>>]
```

Now, if I access the contents attribute, it is loaded from the database:

```
puts news_articles[0].contents
```

The output is:

```
A large tidal wave travelled across the pacific, missing Hawaii by 500 miles
```

Printing the news article using pp news_articles now shows the contents in this output:

```
[#<NewsArticle id=1 url="http://test.com/bigwave" title="Tidal Wave Misses ➥
Hawaii" summary="Tidal wave missed Hawaii by 500 miles" contents="A large ➥
tidal wave travelled across the pacific, missing Hawaii by 500 miles">]
```

The following code snippet demonstrates modifying an object's attribute and persisting the change:

```
news_articles[0].update_attributes(:url => 'http://test.com/bigwave123')
# note: no save call is required
```

You can use this method update to modify one or more attribute values by specifying the new values using a hash. The method update automatically performs a save-to-database operation.

The following code snippets demonstrate object identity. The method first uses the first row in a database table. As you learned from a previous code snippet, the value of news_articles[0] is an object created from the first row of the news_articles table. The following example demonstrates that both objects are the same:

```
news2 = NewsArticle.first
puts "Object equality test: #{news_articles[0] == news2}"
pp news2
```

Here's the output for the equality test and the pretty-printing of news2:

```
Object equality test: true
#<NewsArticle
 title = "Tidal Wave Misses Hawaii",
 summary = "Tidal wave missed Hawaii by 500 miles",
 contents = "A large tidal wave travelled across the pacific, ➥
missing Hawaii by 500 miles",
 id = 1,
 url = "http://test.com/bigwave123">
```

The following examples show how you could add more data to the database and experiment with associations:

```
NewsArticle.new(:url => 'http://test.com/bigfish',
                :title => '100 pound goldfish caught',
                :summary => 'A 100 pound goldfish was caught by Mary Smith',
                :contents => 'A 100 pound goldfish was caught by Mary Smith ➥
using a bamboo fishing pole while fishing with her husband Bob').save
fishnews = NewsArticle.first(:title => '100 pound goldfish caught')

pp fishnews
```

The output for pretty-printing fishnews is:

```
#<NewsArticle
 title = "100 pound goldfish caught",
 summary = "A 100 pound goldfish was caught by Mary Smith",
 contents = "A 100 pound goldfish was caught by Mary Smith ➥
using a bamboo fishing pole while fishing with her husband Bob",
 id = 2,
 url = "http://test.com/bigfish">
```

Notice that contents is loaded. The reason for this is that the news-article object for the second story was created with a single-result query. Lazily loaded attributes are not loaded for multiple-result queries. The following code snippets create two instances of the Person class and associate them with a news story:

```
mary = Person.new(:name => 'Mary Smith')
mary.save
pp mary
```

Notice that the attribute for news_article_id is not defined when you display the contents of the variable mary:

```
#<Person news_article_id = nil, name = "Mary Smith", id = 1>
```

When I add the object mary to the list of people associated with the fish news story, the news_article_id is still not set in mary until the containing object fishnews is saved:

```
fishnews.people << mary
fishnews.save # save required to set news_article_id in object 'mary'
```

If you examine the MySQL database created in this tutorial, you will see that DataMapper created the news_article_id column in both the people and places database tables. If I use the build method to add another person to the news story, then no save is required:

```
fishnews.people.build(:name => 'Bob Smith') # no save required

pp fishnews
pp fishnews.people
```

The output from these three pretty-print statements is:

```
#<NewsArticle
 title = "100 pound goldfish caught",
 summary = "A 100 pound goldfish was caught by Mary Smith",
 contents = "A 100 pound goldfish was caught by Mary Smith ➥
using a bamboo fishing pole while fishing with her husband Bob",
 id = 2,
 url = "http://test.com/bigfish">
[#<Person id=1 name="Mary Smith" news_article_id=2>, ➥
#<Person id=nil name="Bob Smith" news_article_id=2>]
```

Migrating to New Database Schemas

You saw the use of auto-migrations of all DataMapper model classes in the last section:

```
DataMapper.auto_migrate!
```

This will delete the data in all of the database tables for your models. This is not always convenient! You can auto-migrate individual model classes instead, as in this example:

```
Person.auto_migrate!
```

This deletes all data in the people table, changes the table schema to reflect new changes in the Person model class, and unfortunately ruins any associations with data in other model classes. If you are developing Merb-based web applications, then you have another tool to use. You can create a migration rake task:

```
merb-gen migration people_migration_1
```

This allows you to write migrations similar in syntax and function to Rails ActiveRecord migrations. As I am writing this, the Merb and Rails projects are being merged. When this merge is complete, you should be able to use DataMapper in Rails much more easily, and you should be able to take advantage of a common syntax and functionality for migrations. I am not covering the syntax of DataMapper migrations here because they are likely to change.

You obviously do not want to run any auto-migrations that would affect legacy databases.

Using Transactions

There are two ways of handling transactions in DataMapper, but I am only going to show you the easier way: using the model class method `transaction` inside any other methods defined in your model classes. Here is an example that shows the change of a person's name followed by a change back to the original value. Either both database saves have an effect on the database, or neither of them has any effect:

```ruby
class Person
  def transaction_example
    transaction {|a_transaction|
      old_name = self.name
      self.name += ', Ruby master'
      self.save
      self.name = old_name
      self.save
    }
  end
end
```

The method `transaction` is defined by the mix-in `include DataMapper::Resource` when DataMapper model classes are defined.

Modifying Default Behavior

In ActiveRecord you used `set_table_name` to change the default table name calculated from a model-class name. DataMapper also allows you to override the table name:

```ruby
storage_names[:default] = 'existing_table_name'
```

To have a primary key with a name other than `id`, you can use a third argument to the DataMapper `property` method; this argument is a hash of options. The option you set here is the database field (or column) name:

```ruby
property :id, Serial, :field => 'clientID'
```

DataMapper supports composite database keys, so if you wanted keys for first and last names, for example, then you could use something like this:

```ruby
property :first_name, String, :key => true
property :last_name, String, :key => true
```

Handling Callbacks and Observers in DataMapper

Handling callbacks and observers in DataMapper is similar to handling them in ActiveRecord, but the syntax differs. The following example implements an observer class and defines two callback handlers for the class Person (these code snippets are in the file dm_setup_example.rb). In addition to loading dm-core, you need the dm-observer gem:

```
require 'dm-observer'
```

I'll modify the class Person that we used in the DataMapper tutorial to define callbacks that are called after a new person object is created and before an object is saved to the database:

```
class Person
  include DataMapper::Resource
  property :id,   Serial
  property :name, String

  belongs_to :news_article

  # callback examples:
  before :save do
    puts "* * * Person callback: before save  #{self}"
  end
  after :create do
    puts "* * * Person callback: after create  #{self}"
  end
end
```

The class PersonObserver includes (mixes in) the module DataMapper::Observer and uses the methods observe, after, and before that are defined in this module:

```
class PersonObserver
  include DataMapper::Observer

  observe Person

  after :create do
    puts "** PersonObserver: after create #{self}"
  end

  before :save do
    puts "** PersonObserver: before save  #{self}"
  end
end
```

Something that is not obvious, but important, is that the code bodies defined in the before and after method calls are merged into the class being observed. These code blocks print out the instance self, and when you look at output from creating and saving a person object, you see that self prints as a Person instead of a PersonObserver:

```
* * * Person callback: after create  #<Person:0x113cbfc>
** PersonObserver: after create #<Person:0x113cbfc>
* * * Person callback: before save  #<Person:0x113cbfc>
** PersonObserver: before save  #<Person:0x113cbfc>
```

Notice that for the save events, the callbacks execute before the observers.

Wrapup

Relational-database technology is required for most Web 3.0 applications, and you can save a lot of development time using ORM tools. In later chapters, you'll use relational databases for storing data scraped from web sites and for publishing relational data as RDF. I hope this chapter served as a good introduction, especially if you have had little experience with Ruby ORM tools. If you have used ActiveRecord or DataMapper before, I hope you learned something new.

DataMapper shows a lot of promise, but I currently work mostly with ActiveRecord. That said, I think that as DataMapper gets JRuby support in the future, it will probably become my ORM of choice.

Because you can adapt DataMapper to work with nonrelational database sources, you can—with some work—use DataMapper for data in flat files, networked data stores such as CouchDB, or RDF data stores.

Relational databases often support free-text index and search. You'll look at both PostgreSQL's and MySQL's built-in support for text search in Chapter 9. You'll also use the Sphinx, Solr, and Nutch search engines.

CHAPTER 9

■■■

Supporting Indexing and Search

I think that Web 3.0 will feature small, fine-grained, focused applications that encapsulate data and services. These web applications often provide direct search functionality through both web interfaces and remote APIs for searching the published data. I'll explore several approaches for supporting search. I start this chapter with an overview of available off-the-shelf options for indexing and search. I will discuss the pros and cons of these options and the factors that might affect your selection decisions. For reference, I will provide some examples of using each of the search options covered in the first part of this chapter.

Most of the search frameworks covered in this chapter can be used with clients written in a variety of programming languages. I am covering the specific use of Ruby in scripts and applications, but you will find most of the information applicable to other programming languages as well.

I'll begin by using the Java Lucene search library embedded in JRuby applications and scripts. I'll follow this with two Java-based search web services, Solr and Nutch, which you'll use with Ruby clients. A different approach to indexing and search is the Sphinx system, which works by creating external indices to data stored in relational databases. I'll close out this chapter with a tutorial on using PostgreSQL and MySQL internal support for full-text indexing and search.

For the first few years using Ruby, I almost exclusively used the Ferret search library for applications requiring indexing and search. Unfortunately, this fine library is no longer under active development, so I chose not to cover it.

Does your application need to scale over many servers? Sphinx, Solr, and Nutch are good choices if your current application has to scale. In cases where scaling is not as important as ease of use, you'll find it easier to use Lucene embedded in JRuby or the native index and search capabilities of MySQL and PostgreSQL.

Using JRuby and Lucene

Lucene is a best-of-breed indexing and search library written in Java, and it serves as a core part of many other systems. David Andersen wrote a JRuby gem to wrap the Lucene library; you can get it from `http://github.com/davidx/jruby-lucene/tree/master`. Use the `git` utility to fetch the source, then build and install the gem:

```
git clone git://github.com/davidx/jruby-lucene.git
cd jruby-lucene
gem build
gem install jruby-lucene-*.gem
```

You can also add the GitHub site to your gem source list using the following gem command:

```
gem sources -a http://gems.github.com
```

You can then gem install jruby-lucene. I also have this gem prebuilt in the src/part3/ jruby-lucene directory within this book's source code (download it from the Apress web site). You can install the gem from this directory using:

```
gem install jruby-lucene-*.gem
```

To use this gem interactively, try:

```
$ jirb
irb(main):001:0> require 'jruby/lucene'
=> true
irb(main):002:0> lucene = Lucene.new('./temp_data')
=> #<Lucene:0x38ff3be5 @index_path="./temp_data">
irb(main):003:0> lucene.add_documents([[1,"The dog ran quickly"],
                                        [2,'The cat slept in the sunshine']])
=> nil
irb(main):004:0> results = lucene.search('dog')
=> [[0.8465735912322998, 1, "The dog ran quickly"]]
irb(main):005:0> results = lucene.search('The')
=> []
```

In this example, searching for the word "The" yields no results because this word is a *stop word* that is discarded by the standard Lucene tokenizer. When adding documents, you need to supply a unique ID for each document. Supplying an existing ID when creating a new document deletes the previous document. You can also delete documents from the index by specifying their IDs:

```
irb(main):006:0> lucene.delete_documents([1,2])
=> nil
```

If you need to add many documents over time, you might want to keep in a flat file the value for the last document ID added to the index in order to avoid overwriting index entries. The index for a specific directory is persistent. Each time the last code snippet is run, the Lucene index files in the temp_data directory are used.

The next two code snippets implement two command-line scripts for adding text to an index and for searching indexed text. The following indexing script uses an index in the directory ./temp_data and uses a flat file './temp_data/count.txt' to keep a persistent count of the number of documents added to the index so far. The directory temp_data must exist, or the following code snippet will throw an error. This script, which is in the source file index.rb, might fail if multiple copies are run simultaneously:

```ruby
require 'rubygems'
require 'jruby/lucene'

lucene = Lucene.new('./temp_data')
count = 1
begin
  count = File.read('./temp_data/count.txt').strip.to_i
rescue
  puts "Could not open ./temp_data/count.txt"
  File.open('./temp_data/count.txt', 'w') {|f| f.puts('1')}
end

lucene.add_documents(ARGV.collect {|z| count += 1; [count, z]})

File.open('./temp_data/count.txt', 'w') {|f| f.puts(count)}
```

The script search.rb is simpler because you do not have to keep track of document counts:

```ruby
require 'rubygems'
require 'jruby/lucene'
require 'pp'

lucene = Lucene.new('./temp_data')
pp lucene.search(ARGV[0]) if ARGV.length > 0
```

Here is an example use of these two scripts. Each string "document" added to the index is assigned a unique and monotonically increasing document ID. The return value for a search is an array containing a search-relevance value, the document ID, and the original indexed text for the document:

```
$ jruby index.rb "This is a test" "the dog ran"
$ jruby index.rb "The sky is blue" "the cat ran"
$ jruby search.rb "sky"
[[1.0, 5, "The sky is blue"]]
$ jruby search.rb "ran"
[[1.0, 4, "the dog ran"],
 [1.0, 6, "the cat ran"]]
```

The jruby-lucene gem provides a simple-to-use interface to Lucene, and you can easily extend it to use other Lucene features that you might want to use in your applications. You can refer to the Lucene documentation at http://lucene.apache.org. Lucene, widely considered to be a best-of-breed index and search system, is the basis for the extended Solr and Nutch systems that you'll also be using. It's also a great example of the benefits of open source. Because Lucene is so widely used and has such a large community around it, all the details are done right, from the mathematical model for scoring search results, to efficient storage, to high performance for executing queries.

If you need to perform spatial search (for example, to find all businesses within some distance of a location), there is a user-contributed extension to Lucene called Lucene-spatial (http://wiki.apache.org/lucene-java/SpatialSearch). I'll leave it as an exercise for the reader to use this extension with JRuby, but I will show another approach for performing spatial search in Ruby in the next section.

Doing Spatial Search Using Geohash

Geohash is a public-domain algorithm for encoding latitude/longitude positions as a sequence of characters. Geohash has the interesting and useful property that allows shorter substrings starting with the first character of this sequence to also define a position, but with less accuracy. The accuracy is a rough function of how many characters are used.

First I'll show you an example implementation that does *not* use Geohash and that will not scale to handling a very large number of locations to search. I will assume that you have a database of locations you want to search. An example schema containing columns for latitude and longitude positions might be:

```
create table locations (id integer, name varchar(30), lat float, lon float);
insert into locations values (1, 'store1', 39.0234375, -76.552734375);
select * from locations where (lat between 39.0 and 39.5) and
                              (lon between -77.0 and -76.0);
```

This works, but it won't scale to a very large number of locations, even if you create indices for the columns lat and lon. So I'll show you an alternative approach that *does* use the Geohash algorithm. (For details about the algorithm, visit http://en.wikipedia.org/wiki/Geohash.) After implementing both the pure SQL and Geohash approaches, I'll run a benchmark to see which is fastest, and by how much.

I will use Dave Troy's Geohash Ruby gem that you need to install using gem install davetroy-geohash. I'll assume that you have added the GitHub site (http://github.com) to your gem source list.

■**Caution** Currently davetroy-geohash works with Ruby 1.8.6 but not 1.9.x.

I am going to use the following SQL schema for the second implementation and also to benchmark the first pure SQL solution:

```
create table locations (id int, name varchar(30), geohash char(5),
                          lat float, lon float);
create index geohash_index ON locations (geohash, lat, lon);
```

The example files for this section are in the src/part3/spatial-search directory within this book's source code (downloadable from the Apress web site). The Ruby script create_data.rb creates a tiny test database with only 50,000 records:

```
require 'rubygems' # needed for Ruby 1.8.6
require 'activerecord'
require 'geohash' # requires Ruby 1.8.6
require 'pp'
```

Here I establish a connection to the local MySQL database test:

```
ActiveRecord::Base.establish_connection(
  :adapter  => :mysql,
  :database => 'test',
  :username => 'root'
)

class Location < ActiveRecord::Base
end

NUM = 50000 # number of database rows to create with random data
```

The following database inserts should be done in a transaction for larger data sets. But because you're creating only 50,000 rows and because you only need to run this script once, it does not really matter for this example:

```
NUM.times {|n|
  lat =  39.0 + 0.01 * rand(100).to_f
  lon = -77.0 + 0.01 * rand(100).to_f
  geohash = GeoHash.encode(lat, lon)
  Location.new(:name => "location name #{n}", :geohash => geohash[0..4],
    :lat => lat, :lon => lon).save!
}
```

The test database contains latitude and longitude points within a one-degree latitude and a one-degree longitude area. The method GeoHash.encode converts a location to a string geohash. Here are some examples:

```
$ irb
>> require 'rubygems'
=> false
>>   require 'geohash'
=> true
>> GeoHash.encode(39.51, -76.24)
=> "dr1bc0edrj"
```

```
>> GeoHash.encode(39.02, -76.77)
=> "dqcmvp3gcr"
>> GeoHash.encode(38.72, -75.54)
=> "dqfk30c2q7"
```

It might be surprising, but you don't need to use the decoding method; you are only interested in using Geohash encodings to quickly find other "nearby" Geohash values. Comparing more characters between two Geohash values makes the position comparison more accurate. In this example, I am discarding all but the first five Geohash-encoding characters.

The Ruby script file spatial-search-benchmark.rb compares the time required for both algorithms: using pure SQL queries, and using Geohash values. The result that you are calculating is the set of all locations in the database within 0.01 degrees in latitude and longitude of a randomly chosen test location, and the number returned from this set. By comparing the first five Geohash-encoding characters, you find approximately six times the number of locations that you'd find in the 0.01-degree latitude/longitude interval; this is fine because filtering out false hits is a computationally inexpensive operation. As with the first script, I require the gems I need (note that I am also loading the benchmark gem) and open a database connection:

```
require 'rubygems' # needed for Ruby 1.8.6
require 'activerecord'
require 'geohash' # requires Ruby 1.8.6
require 'benchmark'
require 'pp'

ActiveRecord::Base.establish_connection(
  :adapter  => :mysql,
  :database => 'test',
  :username => 'root'
)
```

You use the following two methods to implement the benchmark code for the pure SQL implementation. Note that you are executing a SQL query in ActiveRecord as you did in Chapter 8:

```
def find_near_using_sql lat, lon
  sql = "select * from locations where (lat between ➥
             #{lat - 0.01} and  #{lat + 0.01}) and
                                 (lon between ➥
             #{lon - 0.01} and #{lon + 0.01})"
  ActiveRecord::Base.connection.execute(sql).num_rows
end

def sql_test
  100.times {|n|
    lat =  39.0 + 0.01 * rand(100).to_f
    lon = -77.0 + 0.01 * rand(100).to_f
    find_near_using_sql(lat, lon) # print this value for the number of hits
  }
end
```

The second implementation using Geohash is not as simple because you need to filter out false hits that are outside the 0.01-degree latitude/longitude interval:

```
def find_near_using_geohash lat, lon
  geohash = GeoHash.encode(lat, lon)[0..4]
```

The following SQL query finds all locations that match the first five characters in Geohash encodings:

```
  sql = "select * from locations where geohash = '#{geohash}'"
  count = 0
```

The following code does the (very fast) lookup by Geohash and then discards the false hits that are outside the 0.01-degree latitude/longitude interval. This extra test on latitude and longitude might look inefficient, but compared to the database lookups, it has almost no effect on the runtime:

```
  ActiveRecord::Base.connection.execute(sql).each {|row|
    lat2 = row[3].to_f
    lon2 = row[4].to_f
    count += 1 if lat2 > (lat - 0.01) && lat2 < (lat + 0.01) &&
                        lon2 > (lon - 0.01) && lon2 < (lon + 0.01)
  }
  count
end


def geohash_test
  100.times {|n|
    lat =  39.0 + 0.01 * rand(100).to_f
    lon = -77.0 + 0.01 * rand(100).to_f
    find_near_using_geohash(lat, lon)
  }
end
```

The benchmark code is:

```
puts Benchmark.measure {sql_test}
puts Benchmark.measure {geohash_test}
```

Running the benchmark shows that the second technique (using Geohash) runs about nine times faster than the pure SQL solution:

```
$ ruby spatial-search-benchmark.rb
  0.140000   0.020000   0.160000 (2.867850)
  0.130000   0.000000   0.130000 (0.322963)
```

There are other techniques for using Geohash, but the example in this section will get you started. Using the Wikipedia article and searching for Geohash should provide many interesting examples of different applications that use it.

The next section moves on to a very different subject: using the Solr web services for index and search.

Using Solr Web Services

Solr is a web-service wrapper for Lucene that adds scalability to multiple servers (sharding) and extends the Lucene APIs. You will need to install Solr (http://lucene.apache.org/solr/) to work through the examples in this section. You will also need to install the Apache Tomcat server as per the directions in Chapter 5 for setting up the Sesame web service. The Solr distribution contains a Java WAR file (the file name is apache-solr-1.3.0.war in the latest version of Solr that I have installed) that needs to be copied to the webapps directory in your Tomcat setup. Rename the WAR file in webapps to solr.war.

Copy the example/solr directory from the Solr distribution directory to the Tomcat directory. You'll use the ruby-solr gem (http://wiki.apache.org/solr/solr-ruby), so install it using the command gem install solr-ruby.

From the Tomcat directory, interactively run the Tomcat service:

```
bin/catalina.sh run
```

From the src/part3/solr directory in this book's source code (downloadable from the Apress web site), try running this test client program to make sure that your Solr installation is working:

```
ruby solr_test.rb
```

You should see the following output that shows the additional fields sku and popularity defined in the Solr example configuration (you'll get rid of these later when you modify the configuration):

```
$ jruby solr_test.rb
{"id"=>"2",
 "popularity"=>0,
 "sku"=>"2",
 "timestamp"=>"2009-03-01T15:42:00.2Z",
 "score"=>0.4332343}
{"id"=>"1",
 "popularity"=>0,
 "sku"=>"1",
 "timestamp"=>"2009-03-01T15:41:59.636Z",
 "score"=>0.37908003}
```

The following code snippets are in the file solr_test.rb. This example starts with making a connection to the Solr web service:

```
require 'rubygems'
require 'solr'
require 'pp'

solr_connection = Solr::Connection.new('http://localhost:8080/solr',
                                        :autocommit => :on)
```

When you use Solr's web services, you use HTML POSTs containing XML requests to the web service. The `solr-ruby` gem simplifies things considerably, at the expense of not having the full functionality of Solr. In this example, I use the add method to define new documents:

```
solr_connection.add(:id => 1, :text => 'The dog chased the cat up the tree')
solr_connection.add(:id => 2, :text => 'The cat enjoyed sitting in the tree')
```

The document field `:text` is a default Solr field to index and search. I could also have added values for `:popularity` and `:sku` that are specific to the Solr example configuration, but in the example I specify only text to add to the `:text` field.

The method query encodes a search string in an XML request and sends it to the remote Solr web service. In this example, I am pretty-printing the results for each search result (the variable `hit`):

```
solr_connection.query('tree') {|hit| pp hit}
```

The output for this query is:

```
{"score"=>0.5, "timestamp"=>"2009-03-31T18:14:47.155Z", "id"=>"2"}
{"score"=>0.4375, "timestamp"=>"2009-03-31T18:14:46.894Z", "id"=>"1"}
```

You can also update the value of a document field in the index:

```
solr_connection.update(:id => 1, :text => 'The dog went home to eat')
```

You can delete index entries by specifying their IDs:

```
solr_connection.delete(1)
solr_connection.delete(2)
```

The keys id, timestamp, and score are expected in search: you need to identify a matched document resource (id), you want to know when a matched document was added to the index (timestamp), and you want the ranking (score) of the result. The other result keys, popularity and sku, are artifacts of the example Solr configuration that you copied to Tomcat/solr earlier. The example bundled with the Solr distribution is set up for searching for products; sku is a product identifier and popularity indicates how many people bought a given product.

I backed up the Tomcat/solr/conf/solrconfig.xml and edited it to delete the configuration for spelling correction, which is not presently available for solr-ruby. I also backed up Tomcat/solr/conf/schema.xml and edited it to eliminate the example schema for searching for a manufacturer's products. My edited solrconfig.xml and schema.xml files are in the src/part3/solr directory within the book's source code. If you copy these files to your Tomcat/solr/conf directory (overwriting the standard Solr example files) and restart Tomcat, then your search query results will look like this:

```
{"id"=>"2", "timestamp"=>"2009-03-01T17:16:45.749Z", "score"=>0.2972674}
{"id"=>"1", "timestamp"=>"2009-03-01T17:16:45.215Z", "score"=>0.26010898}
```

To monitor the Solr search-engine statistics running in a Tomcat container, use this URL:

```
http://localhost:8080/solr/admin/stats.jsp
```

Solr has additional functionality like spelling correction that is documented on the Solr web site. Exploring the full functionality of Solr is way outside the scope of this book, but following the examples in this section should prepare you to evaluate Solr for your projects. In the next section, you'll study another search system called Nutch that is built using Lucene. It provides a complete and scalable "Google in a box" system that is simple to configure and deploy.

Using Nutch with Ruby Clients

The Nutch system is an older project than Solr. Whereas Solr is a component for building systems, Nutch is a complete "turnkey" web-portal solution for search functionality. Nutch, when you use its default options, is very simple to configure and deploy. It contains a "spidering" module that can index web pages, follow web-page links to a specified depth and index those pages, and index a variety of document formats such as Microsoft Office, PDF, and so on.

I am going to provide you with a brief tutorial for configuring and deploying Nutch in this section so that you can evaluate the use of Nutch for your projects. You can choose from two approaches when it comes to using Nutch: you can use it as a web service through a Ruby client, or you can use Nutch's built-in web interface. The latter doesn't require much explanation, so I'll devote the bulk of this section to the former approach.

Nutch uses the OpenSearch web-service interface standard, so any client code that you write to work with Nutch can be used for other OpenSearch-compliant search engines. The Nutch web site (http://lucene.apache.org/nutch) features distributions and tutorials for quickly setting up a Nutch system. As an example, I will configure Nutch to periodically "spider" two of my own web sites and then use a web-service client written in Ruby to submit queries.

To install Nutch, you need to download and install Tomcat as you did in the last section and in Chapter 5. Delete all files and directories in Tomcat/webapps and then move the Nutch WAR file to the file Tomcat/webapps/ROOT.war (that is, copy it and change its name so it will install as the root Tomcat context). After moving the WAR file, change the name of the Nutch distribution directory to nutch and move this directory to the top-level Tomcat directory. The Tomcat directory structure should now look like the following (only directories and files of interest are shown here):

```
Tomcat
Tomcat/bin
Tomcat/nutch
Tomcat/nutch/conf
Tomcat/nutch/conf/crawl-urlfilter.txt
```

```
Tomcat/nutch/conf/nutch-site.xml
Tomcat/nutch/urls/starting_urls
Tomcat/webapps
Tomcat/webapps/ROOT.war
```

Whenever you set up a new Nutch system, you need to edit the configuration files
`crawl-urlfilter.txt` and `nutch-site.xml`. I have edited versions of these files in the `src/`
`part3/nutch` directory (downloadable from the Apress web site); copy them to your `Tomcat/`
`nutch/conf` directory, overwriting the existing files. You then need to edit `nutch-site.xml` and
replace `YOUR_NAME`, `YOUR_DOMAIN`, and `YOUR_EMAIL_NAME` with appropriate values to identify yourself to any sites that you are spidering to build a local index. My version of `crawl-urlfilter.`
`txt` configures Nutch to spider only web pages at my web sites `http://markwatson.com` and
`http://knowledgebooks.com`. Here's the snippet containing regular expressions matching my
domains from `crawl-urlfilter.txt`:

```
# accept hosts in markwatson.com
+^http://([a-z0-9]*\.)*markwatson.com/

# accept hosts in knowledgebooks.com
+^http://([a-z0-9]*\.)*knowledgebooks.com/
```

I also modified the file-exclusion list in the crawl filter to exclude PDF files. You will eventually want to change the web sites that your local Nutch system spiders. You need to create a
`Tomcat/nutch/urls` directory that has one or more text files containing seeds (or starting URLs)
that the Nutch spider uses to start following links and indexing the web pages that it finds. I
created a file called /`Tomcat/nutch/urls/seeds` that contains these seeds:

```
http://markwatson.com
http://knowledgebooks.com
```

You will want to modify this for the sites matching your crawl filters. To run a test spider
operation (which might take a few minutes) and then start the Tomcat server, use the following commands:

```
cd Tomcat/nutch/
rm -f -r crawl
bin/nutch crawl urls -dir crawl -depth 2 -topN 100
../bin/catalina.sh run
```

Note that for each search level, I set the spider search depth to 2 (that is, to follow up to
two links from the seed URLs), and I limited the number of spidered pages to 100.

It is important for you to run the preceding command to start Tomcat from inside the
`Tomcat/nutch` directory and not from the top-level `Tomcat` directory. Figures 9-1 and 9-2 show
the default Nutch web application that you can access using `http://localhost:8080` if you
installed Nutch on your local laptop or workstation.

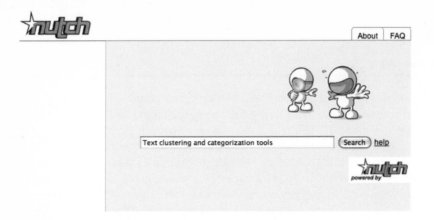

Figure 9-1. *Nutch web application showing a search query*

You can change the graphics and layout for the application by editing the files in the directory `Tomcat/webapps/ROOT`. Figure 9-2 shows example search results. If you have installed Nutch and are following along with this example, click the search-results link "explain" to see how search results are scored.

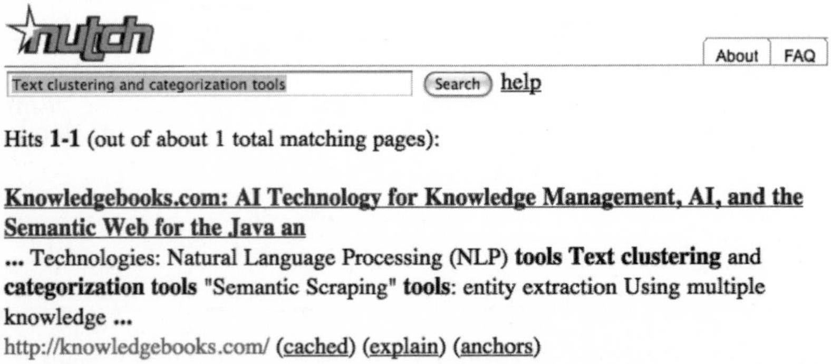

Figure 9-2. *Nutch web application showing search results*

The OpenSearch web-service APIs are REST queries similar to what you used when you wrote Ruby clients for the Sesame web-service APIs. The root search URL is:

```
http://localhost:8080/opensearch?query=
```

Replace `localhost` with the domain name where you installed Nutch if you are not testing locally. All you need to do is to URL-encode a query, append it to the root search URL, and then perform an HTTP `GET` operation. Here is a Ruby test script that performs a REST search request by using the `restclient` gem and pretty-prints the XML search results:

```
require 'rubygems'
require 'cgi'
```

```
require 'rexml/document'
include REXML

require 'restclient'

ROOT_SEARCH_URL = 'http://localhost:8080/opensearch?query='
query = 'text clustering categorization tools'

url = ROOT_SEARCH_URL + CGI.escape(query)
results = RestClient.get(url)
xml_doc = Document.new(results)
xml_doc.write($stdout, 0)
```

The output from this Ruby script is shown in the next snippet, with long lines broken with a code-continuation character (➡) and shown on multiple lines. The description element shows only part of the URL-encoded HTTP text for displaying search results:

```
<?xml version='1.0' encoding='UTF-8'?>

<rss xmlns:nutch='http://www.nutch.org/opensearchrss/1.0/' version='2.0'
        xmlns:opensearch='http://a9.com/-/spec/opensearchrss/1.0/'>
  <channel>
    <title>Nutch: text clustering categorization tools</title>
    <description>
      Nutch search results for query: text clustering categorization tools
    </description>
    <link>http://localhost:8080/search.jsp?query= ➡
                  text+clustering+categorization+tools ➡
                  &amp;start=0&amp;hitsPerDup=2&amp;hitsPerPage=10
    </link>
    <opensearch:totalResults>1</opensearch:totalResults>
    <opensearch:startIndex>0</opensearch:startIndex>
    <opensearch:itemsPerPage>10</opensearch:itemsPerPage>
    <nutch:query>text clustering categorization tools</nutch:query>
    <item>
      <title>Knowledgebooks.com: AI Technology for Knowledge Management, AI, ➡
              and the Semantic Web for the Java, Ruby, and Common Lisp Platforms
      </title>
      <description>&lt;span class="ellipsis"&gt; ... &lt;/span&gt;Technologies: ➡
                      Natural Language Processing (NLP) &lt;span ➡
                          ...
       </description>
      <link>http://knowledgebooks.com/</link>
      <nutch:site>knowledgebooks.com</nutch:site>
      <nutch:cache>http://localhost:8080/cached.jsp?idx=0&amp;id=0</nutch:cache>
      <nutch:explain>
          http://localhost:8080/explain.jsp?idx=0&amp;id=0&amp;query= ➡
                  text+clustering+categorization+tools&amp;lang=null
```

```
        </nutch:explain>
        <nutch:segment>20090301121248</nutch:segment>
        <nutch:digest>923fb80f9f8fd66f47d76d1ab55c8ea0</nutch:digest>
        <nutch:tstamp>20090301191253937</nutch:tstamp>
        <nutch:boost>1.0</nutch:boost>
      </item>
    </channel>
</rss>
```

The XML response has an element `description` that contains URL-encoded HTML for displaying a search result. This is useful when your search client is another web application that uses Nutch search. The element `link` contains the URI for the result document.

For many applications, you might choose the approach you used in this section: to use Nutch as a web service via a client. But the out-of-the-box web interface for Nutch is full-featured, and you might simply choose to integrate the Nutch web interface into your own web applications. With some customization of the HTML and CSS of the Nutch web interface to match the look of your web application, you can smoothly integrate Nutch into larger web portals that require search capability.

We move on to a different topic in the next three sections, where you'll see three techniques for efficiently searching for data inside relational databases.

Using Sphinx with the Thinking Sphinx Rails Plugin

I'll close out this chapter with three techniques for efficiently searching for information inside relational databases. I think that the most powerful technique is to use the external Sphinx indexing and search service; I cover Sphinx in this section. The next two sections cover built-in search capabilities in PostgreSQL and MySQL, which are less efficient but easier to implement, configure, and deploy. In other words, the search functionality in PostgreSQL and MySQL suffices for situations that do not require Sphinx's higher level of efficiency and scalability.

Sphinx supports native interfaces to both MySQL and PostgreSQL and has client libraries implemented in Ruby, Python, Perl, PHP, and Java. The example in this section will use MySQL with the high-level Ruby access library Thinking Sphinx (`http://ts.freelancing-gods.com/`).

Because Thinking Sphinx is a Ruby on Rails plugin, the example program in this section is a small Ruby on Rails web application. I assume that you have some familiarity with Ruby on Rails, both here and in Chapter 15, where I'll develop two more Rails applications. If you have never used Ruby on Rails, please take a half hour to read through the introductory material on the Rails web site (`http://rubyonrails.org/`). I am using Rails version 2.3 for the examples in this chapter.

Installing Sphinx

You start by installing and configuring Sphinx (`http://www.sphinxsearch.com`). You can refer to the documentation at the Sphinx web site, but for my Linux servers and my MacBook I simply fetched the latest source-code distribution file build and installed it from source code:

```
tar xzvf sphinx-0.9.8.tar.gz
cd sphinx-0.9.8
./configure --prefix=/usr/local/sphinx
make ; sudo make install
```

If the Sphinx build process cannot find your MySQL and PostgreSQL installation direc-tories, you might also have to use the --with-mysql and --with-pgsql configure options. You don't need to do any additional Sphinx configuration except for making sure that the Sphinx bin directory is on your path. The Ruby on Rails Thinking Sphinx plugin adds Rake options for configuring and managing Sphinx.

Note I have this example web application set up and running on the Amazon Machine Image (AMI) that I created for all of the examples in this book (see Appendix A). If you use my AMI, you will find this application running on port 3011; in this section, though, I use the default Rails port 3000.

Installing Thinking Sphinx

The Thinking Sphinx plugin installs into a Rails application (the finished example web appli-cation is in src/part3/thinking-sphinx-rails-demo). You first need to install the chronic gem using gem install chronic. Because you might be installing Thinking Sphinx in your own projects, I'll take some time to go through the steps you will need to follow for your own proj-ects. I start by creating an empty Rails project:

```
rails thinking-sphinx-rails-demo
cd thinking-sphinx-rails-demo
script/plugin install git://github.com/freelancing-god/thinking-sphinx.git
```

After setting up my config/database.yml file to use MySQL, the MySQL database test, and the news_articles table used in Chapter 8, I can generate a model:

```
script/generate model NewsArticle
```

As you might remember from Chapter 8, the table news_articles has columns id, url, title, summary, and contents. For this example, I am going to have Sphinx index the columns title and summary by editing the generated model file and adding a call to the method define_index that is provided by the Thinking Sphinx plugin:

```
class NewsArticle < ActiveRecord::Base
  define_index do
    indexes [:title, :contents]
  end
end
```

Then I run rake tasks to create index files for Sphinx and start the Sphinx process as a daemon:

```
$rake thinking_sphinx:index
(in /home/markw/Documents/WORK/ruby_scripting_book/src/part3/➡
thinking-sphinx-rails-demo)

Generating Configuration to /home/markw/Documents/WORK/ruby_scripting_book/src/➡
part3/thinking-sphinx-rails-demo/config/development.sphinx.conf

indexer --config /home/markw/Documents/WORK/ruby_scripting_book/src/part3/➡
thinking-sphinx-rails-demo/config/development.sphinx.conf --all

Sphinx 0.9.8.1-release (r1533)

Copyright (c) 2001-2008, Andrew Aksyonoff


using config file '/home/markw/Documents/WORK/ruby_scripting_book/src/part3/➡
thinking-sphinx-rails-demo/config/development.sphinx.conf'...

indexing index 'news_article_core'...

collected 1 docs, 0.0 MB

collected 0 attr values

sorted 0.0 Mvalues, 100.0% done

sorted 0.0 Mhits, 100.0% done

total 1 docs, 101 bytes

total 0.010 sec, 10100.00 bytes/sec, 100.00 docs/sec

distributed index 'news_article' can not be directly indexed; skipping.


$ rake thinking_sphinx:start
(in /home/markw/Documents/WORK/ruby_scripting_book/src/part3/➡
thinking-sphinx-rails-demo)

searchd --pidfile --config /home/markw/Documents/WORK/ruby_scripting_book/➡
src/part3/thinking-sphinx-rails-demo/config/development.sphinx.conf
```

```
Sphinx 0.9.8.1-release (r1533)

Copyright (c) 2001-2008, Andrew Aksyonoff
```

```
using config file '/home/markw/Documents/WORK/ruby_scripting_book/src/part3/➥
thinking-sphinx-rails-demo/config/development.sphinx.conf'...

creating server socket on 127.0.0.1:3312

Started successfully (pid 17898).
```

As you have just seen, configuring and running Sphinx through Thinking Sphinx is simple. If you need to, you can configure Sphinx manually and then follow the directions at http://www.sphinxsearch.com/docs. In order to make this demo web application complete, I am going to add a controller and view for the default (index) web page:

```
script/generate controller Index
```

I edit the config/routes.rb file to add a default route to the controller index:

```
map.root :controller => "index"
```

You remember that I defined the attributes title and contents of the NewsArticle class as searchable in the model class definition, so I can use the method search in the index controller:

```
class IndexController < ApplicationController
  def index
    pp params
    @results = []
    @query = params['search_text']
    if @query
      @results = NewsArticle.search(@query)
      pp @results
    else
      @query = ''
    end
  end
end
```

I wanted to keep the old query string in @query so that the view can show the user the previous query string when the default web page refreshes. The array @results contains instances of the class NewsArticle that match the query. The following snippet shows some of the view file app/views/index/index.rhtml:

```
<form action="/" method="get">
  <input type="text" size="30" name="search_text" value="<%=@query%>"/>
  <input type="submit" value="Search"/>
</form>

<% @results.each {|result| %>
  <p>Matching title: <%= result.title %></p>
<% } %>
```

Rails provides helper methods for generating forms and other HTML elements, but I used a standard HTML form element in the preceding code snippet, so you can still understand this example even if you are not experienced with Rails. Rails renders view files by replacing any <%= %> brackets with the value of the expression inside the brackets. Any Ruby code inside <% %> brackets is executed but not rendered into generated HTML. Figure 9-3 shows this demo search web application.

Thinking Sphinx Search Demo

Enter Search request:

| wave | Search |

Matching title: Tidal Wave Misses Hawaii

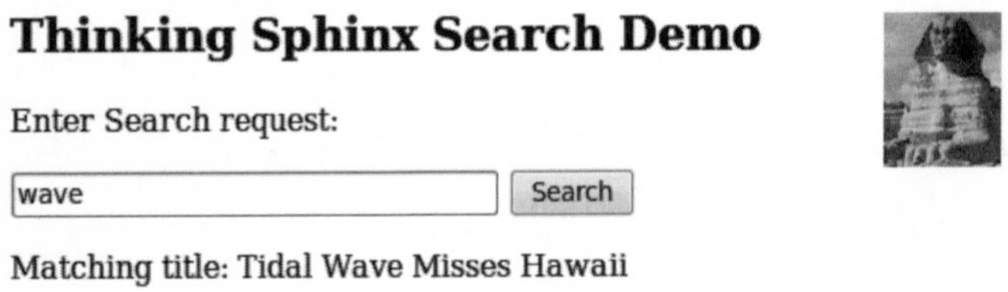

Figure 9-3. *Demo search web application using Sphinx and Thinking Sphinx*

Sphinx is a scalable search solution that, for example, powers the http://www.craigslist. org web sites. In general, setting up a Sphinx deployment is not simple, but because Thinking Sphinx does most of the work of configuring and running Sphinx for you, it is easy for you to experiment with and use Sphinx in your own Rails applications.

Thinking Sphinx has some cool features that I have not used in this simple example. You can find the reference and setup instructions for Thinking Sphinx at http://ts.freelancing-gods.com/ and http://ts.freelancing-gods.com/usage.html.

The next two sections cover the use of the built-in, full-text search capabilities of PostgreSQL and MySQL.

Using PostgreSQL Full-Text Search

Starting with the release of version 8.3, PostgreSQL supports full-text indexing and search of selected columns in database tables. PostgreSQL is my favorite database to use for projects, and now that it contains full-text indexing and search, I often use it as a "complete" database and search engine solution.

I'll start this section with sample interactive SQL queries demonstrating index and search. There are two Ruby libraries you could use: the TSearchable Ruby plugin library at http://github.com/dylanz/tsearchable, and the acts_as_tsearch project at http://code.google. com/p/acts-as-tsearch. I chose to not use either of these libraries, though; instead, I wrote my own monkey patches to the ActiveRecord base class to support PostgreSQL full-text search. At the end of this chapter, I'll show you monkey patches for MySQL full-text search that use the same method signatures, so you can have some portability for switching between PostgreSQL and MySQL.

My first preference is to avoid using tools that lock me into one technology. One reason for using ORM tools like ActiveRecord and DataMapper is the ability to change underlying database systems (see Chapter 8). In this section, you'll use SQL and database extensions specifically for PostgreSQL. However, PostgreSQL is open source, it's supported by third-party vendors if you need to purchase support, and it happens to be my favorite database system.

PostgreSQL can scale to multiple servers, and if properly tuned (for example, if you configure PostgreSQL services with sufficient memory for effective caching), it provides good performance for data insertion, indexing, queries, and free-text searching.

The PostgreSQL-specific SQL extensions for indexing and search might take some getting used to. The following is an interactive session that creates a new database, creates a table with indexing enabled on the column contents, and performs example free-text searches:

```
$ createdb search_test -U postgres
$ psql search_test -U postgres
Welcome to psql 8.3.4, the PostgreSQL interactive terminal.
search_test=# create table articles (id integer, title varchar(30), ➥
contents varchar(250));
CREATE TABLE
```

In the next snippet, you'll create an index on the column contents using a Generalized Inverted iNdex (GIN). The alternative type of index is Generalized Search Tree (GiST), which uses balanced trees. GIN is an inverted word hash with words and word stems as keys, and collections of row numbers as hash values. GiST uses word signatures that are combined with logical-OR operations. GiST produces false search hits that PostgreSQL automatically filters out by checking the actual data rows. GIN provides faster queries while GiST provides faster indexing. You can benchmark both index types using representative data for your application, but as a rule of thumb, use GIN when database data is relatively static and use GiST when you are frequently adding and updating data.

The function to_tsvector parses text into tokens, word-stems these tokens, and returns a list of word stems for the input text. In this example, I'm stating that the text data in the column contents is expected to be English text:

```
search_test=# create index articles_contents_idx on articles
                              using gin(to_tsvector('english', contents));
CREATE INDEX
search_test=# insert into articles values (1, 'Fishing Season Open',
                              'Last Saturday was the opening of Fishing season');
INSERT 0 1
search_test=# insert into articles values (2, 'Tennis Open Cancelled',
                          'The tennis open last weekend was cancelled due to rain');
INSERT 0 1
```

Using the two test rows in the table articles, I'll show you some test queries. In all of them, the function to_tsquery converts query terms into an internal representation used by PostgreSQL. The next two queries demonstrate how search matches occur for stemmed words. For example, the word "open," in a stemmed form, appears in both rows of the table:

```
search_test=# select id, title from articles
                           where to_tsvector(contents) @@ to_tsquery('open');
 id |          title
----+----------------------
  1 | Fishing Season Open
  2 | Tennis Open Cancelled
(2 rows)
```

And here, "fish" matches "fishing":

```
search_test=# select id, title from articles
                           where to_tsvector(contents) @@ to_tsquery('fish');
 id |         title
----+---------------------
  1 | Fishing Season Open
(1 row)
```

The next query illustrates the use of the "and" operator (&). It matches rows where the column contents matches stems for both "fish" and "salmon," yielding no results:

```
search_test=# select id, title from articles
                          where to_tsvector(contents) @@ to_tsquery('fish & salmon');
 id | title
----+-------
(0 rows)
```

Now you'll see the "or" operator (|) in action. This query matches rows where the column contents matches stems for either "fish" or "salmon," yielding one result:

```
search_test=# select id, title from articles
                          where to_tsvector(contents) @@ to_tsquery('fish | salmon');
 id |         title
----+---------------------
  1 | Fishing Season Open
(1 row)
```

You can also use the "not" operator (!) to specify that matches should be discarded when a word appears. In the next example, the column contents must contain a word stem of "open" and must *not* contain a word stem of "fish":

```
search_test=# select id, title from articles
                      where to_tsvector(contents) @@ to_tsquery('open & !fish');
    id |         title
 ----+----------------------
     2 | Tennis Open Cancelled
(1 row)
```

In the next section, you'll develop a Ruby client script for using PostgreSQL full-text search.

Developing a Ruby Client Script

This section contains an example of using a direct connector to a local PostgreSQL database. In the next section, I will extend ActiveRecord to support full-text search. You can use either the pure Ruby gem postgres-pr, which works with JRuby, or the gem postgres, which requires native extensions and the include files and libraries in your PostgreSQL installation. The following script, which resides in the file src/part3/postgresql-search/pure-ruby-test.rb, uses the pure Ruby PostgreSQL connector gem:

```
require 'postgres-pr/connection'
require 'pp'
```

The two required arguments for creating a connection to PostgreSQL are the name of the database and the name of the account that has access to the database:

```
conn = PostgresPR::Connection.new('search_test', 'postgres')
```

You use the query method for queries, updates, and deletes. Here I pass in a SQL query using the full-text search syntax that you saw in the last section. Each result is an array of query values:

```
results = conn.query(
        "select id, title from articles where to_tsvector(contents) @@ ➥
to_tsquery('fish')")
results.rows.each {|result| pp result}

conn.query("insert into articles values (3, 'Watson Wins Championship', ➥
'Mark Watson won the ping pong championship for the second year in a row')")

results = conn.query("select id, title from articles where ➥
to_tsvector(contents) @@ to_tsquery('ping & pong')")

pp results.rows

# get rid of the new row I just added:
conn.query('delete from articles where id=3')

# close the socket connection to PostgreSQL:
conn.close
```

The results method rows returns an array of arrays. Each subarray contains the results for one row. The output from this example script is:

```
$ ruby pure-ruby-test.rb
["1", "Fishing Season Open"]
[["3", "Watson Wins Championship"]]
```

While it is fine to use native SQL queries for full-text search, I prefer to use ActiveRecord. The example in the next section extends the ActiveRecord base class to support full-text queries.

Integrating PostgreSQL Text Search with ActiveRecord

I mentioned two Ruby libraries at the beginning of the discussion about PostgreSQL: TSearchable and acts_as_tsearch. You could use either of these Rails-oriented projects to add PostgreSQL text search to ActiveRecord, but they do more than what I usually need for my work. So I wrote my own wrapper. In this section, I'll offer two implementations for monkey-patching the ActiveRecord::Base class to use PostgreSQL text search.

My first implementation requires that you provide the name of an indexed text field to search. Our goal is to allow search using something like to_tsquery('fish') to search for the word "fish" in the column contents:

```
Article.text_search_by_column('contents', 'fish')
```

It is simple to implement the ActiveRecord API by generating a SQL query string containing the query terms and returning an array of subarray results:

```
class ActiveRecord::Base
  def self.text_search_by_column column_name, query
    sql =<<-SQL
              select * from #{self.table_name}
          where to_tsvector(#{column_name}) @@ to_tsquery('#{query}')
      SQL
      ActiveRecord::Base.connection.execute(sql).rows
  end
end
```

This code generates an appropriate SQL query, calls the ActiveRecord API method execute for running native SQL queries, and returns the results in an array of subarray string values. For a query for "fish" on the contents of the column contents, the generated SQL query looks like this:

```
select * from articles where to_tsvector(contents) @@ to_tsquery('fish')
```

Here is an interactive irb session using this monkey patch:

```
$ irb
>> require 'postgresql-activerecord-simple'
```

```
>> Article.text_search_by_column('contents', 'fish')
=> [["1", "Fishing Season Open", "Last Saturday was the opening of Fishing season"]]
>> Article.text_search_by_column('contents', 'season | tennis')
=> [["1", "Fishing Season Open", "Last Saturday was the opening of Fishing ➥
season"], ["2", "Tennis Open Cancelled", "The tennis open last weekend was ➥
cancelled due to rain"]]
```

This first implementation has one problem, making it awkward to use: you need to specify a searchable table column in the API calls. For my use, I would prefer that all searchable columns were searched automatically.

In order to implement the second version of my monkey patch to ActiveRecord::Base, I need to test to see which columns are searchable. I was unable to find any direct APIs to determine if a column is full-text–searchable, so instead I'll perform a test text search on each column and collect the column names for which there is no runtime error. This process needs to be done only once. Add the following private class method to ActiveRecord::Base:

```
def self.get_searchable_columns
  ret = []
  self.column_names.each {|column_name|
    begin
      sql =<<-SQL
              select * from #{self.table_name}
          where to_tsvector(#{column_name}) @@ to_tsquery('fish')
          limit 1
      SQL
      ActiveRecord::Base.connection.execute(sql).rows
      # the following statement only gets executed if
      # there were no errors
      ret << column_name
    rescue # ignore this rescue clause
    end
  }
  ret.join(' || ')
end
```

I want the overhead of determining the searchable columns to occur only once, so my new search API only calls the class method get_searchable_columns the first time that it is called:

```
def self.text_search query
  @@searchable_columns ||= get_searchable_columns
  sql =<<-SQL
          select * from #{self.table_name}
      where to_tsvector(#{@@searchable_columns}) @@
          to_tsquery('#{query}')
  SQL
  ActiveRecord::Base.connection.execute(sql).rows
end
```

This is similar to the implementation of the text_search_by_column method, except that you search on all columns containing text data. The select statement looks like this:

```
select * from articles where to_tsvector(title || contents) @@ to_tsquery('fish')
```

Here is an interactive irb session using the second implementation:

```
>> require 'postgresql-activerecord'
>> Article.text_search('fish')
=> [["1", "Fishing Season Open", "Last Saturday was the opening of Fishing season"]]
>> Article.text_search('tennis')
=> [["2", "Tennis Open Cancelled", "The tennis open last weekend was cancelled ➥
due to rain"]]
>> Article.text_search('tennis & !cancelled')
=> []
```

My monkey patches to the ActiveRecord base class provide a convenient and efficient way to perform full-text search on text data in PostgreSQL databases. In the next section, you will see the MySQL extensions for full-text search, and I will provide monkey patches to the Active-Record base class that use the same APIs that I developed in this section.

Using MySQL Full-Text Search

Like PostgreSQL, newer versions of MySQL natively support indexing and search of text stored in database table columns. Also, as with PostgreSQL built-in text search, you don't need to install extra software packages or run other processes.

Full-text search works only on MySQL MyISAM table types. Because this is not the default back-end storage engine, you need to specify the background engine when creating tables. For the example in this section, we will use:

```
create table news (id int, title varchar(30), contents varchar(200)) engine = ➥
MYISAM;
```

MySQL supports a wide variety of query modes for text search, and I will cover the ones that you are most likely to need in the following examples. For a complete reference, see the MySQL documentation page for full-text search (http://dev.mysql.com/doc/refman/5.1/en/fulltext-search.html). If you need to add a lot of text to a database and create a full-text index, it is best to add rows to a table and then index it. In this example, I am adding just three rows and then indexing the table. Any rows added later will then be indexed automatically. Note that I use the keyword fulltext when creating a searchable text index:

```
insert into news values (1, 'Home Farming News', 'Both government officials ➥
and organic food activists agree that promoting home and community gardens ➥
is a first line of defense during national emergencies');
insert into news values (2, 'Families using less and enjoying themselves more', ➥
```

```
'Recent studies have shown that families who work together to grow food, ➡
cook together, and maintain their own homes are 215% happier than families ➡
stuck in a "consumption rut".');
insert into news values (3, 'Benefits of Organic Food', 'There is now more ➡
evidence that families who eat mostly organic food may have fewer ➡
long term health problems.');

create fulltext index news_index on news (contents);
```

I will show you the syntax for full-text search in the next section and then monkey-patch ActiveRecord to support full-text search.

Using MySQL SQL Full-Text Functions

Just as PostgreSQL had its own odd syntax for full-text search, MySQL has its own and different syntax. The two new functions you'll use are match and against. The argument to match is one or more column names that have a full-text index; the arguments to against are search terms enclosed in single quotes followed by optional directions. In the next two examples I'll use boolean search mode, which supports placing plus signs in front of words that you want to see in search results and minus signs in front of words that you don't want to see in search results. One of the example records has the words "grow" and "food" in the contents column, but if you specify "-food" in the query string, then you get an empty result set (long queries are split over two lines for better readability):

```
mysql> select title from news where match (contents)
                                 against ('+grow -food' in boolean mode);
Empty set (0.00 sec)
```

The following query finds both rows that contain the word "organic" in the contents column:

```
mysql> select title from news where match (contents)
                                 against ('+organic' in boolean mode);
+--------------------------+
| title                    |
+--------------------------+
| Home Farming News        |
| Benefits of Organic Food |
+--------------------------+
2 rows in set (0.00 sec)
```

I almost always use MySQL full-text search in boolean mode. In boolean mode, the + operator means "and," the – operator means "not," and there's no operator that means "or."

MySQL full-text search has an interesting and sometimes useful search option called with query expansion. It first makes a search query, and from the top-ranked rows found in the first query, it combines the words in the indexed column(s) with the original query words and performs a second search. This can, for example, correct spelling mistakes in search queries. Query expansion can also allow matching words associated with query terms. As an example, if the original query terms are "president of the United States," then matched rows probably contain "Barack Obama." So the second, expanded query is likely to pick up extra rows that contain "Barack Obama" but not, for example, "president."

As an example of spelling correction, I am going to add a fourth row to our test table that contains the misspelling "heelth" (for "health"):

```
insert into news values (4, 'Benefits of Healthy Food', 'There is now more ➡
evidence that families who eat mostly healthy food may have fewer ➡
long term heelth problems.');
```

If I now search for documents containing "health problems," I get only one result because the row I just added cannot match "health" due to the misspelling:

```
mysql> select title from news where match (contents) against ('health problems');
+-------------------------+
| title                   |
+-------------------------+
| Benefits of Organic Food |
+-------------------------+
1 row in set (0.00 sec)
```

Now, if I use the with query expansion search option:

```
mysql> select title from news where match (contents) against (➡
'health problems' with query expansion);
```

I will find the last row I added:

```
+-------------------------+
| title                   |
+-------------------------+
| Benefits of Organic Food |
| Benefits of Healthy Food |
+-------------------------+
2 rows in set (0.00 sec)
```

The reason this works becomes clear if you look at the words for the first match (for the title "Benefits of Organic Food"):

```
"There is now more  evidence that families who eat mostly organic food may have ➡
fewer long term health problems."
```

Compare the preceding text sample with the words in the contents column for the second row that was matched using the with query expansion search option:

"There is now more **evidence** that **families** who **eat** mostly healthy **food** may have ➥
fewer long **term health problems**."

MySQL full-text search discards *noise words* (or *stop words*), so in the preceding listings of the contents from the two matched rows, I highlighted in bold font the words that are in common and that are not noise words. MySQL full-text search has a feature that I find useful: if a search term appears in at least 50 percent of the rows in a table, then the search term is treated as a noise word and is not used in the query. This might surprise you, however, if you search for a very common word in a database table and get no results.

MySQL full-text search also supports the query option in natural language mode, which is also the default mode. The following two queries are the same:

```
select title from news where match (contents)
                                     against ('health' in natural language mode);
select title from news where match (contents) against ('health');
```

If you do not want to discard results for terms occurring in 50 percent or more of the rows, then use the boolean search mode. Compare these results:

```
mysql> select title from news where match (contents) against ('food');
Empty set (0.00 sec)

mysql> select title from news where match (contents) against ('+food' in ➥
boolean mode);
+------------------------------+
| title                        |
+------------------------------+
| Home Farming News            |
| Families using less and enjoyi |
| Benefits of Organic Food     |
| Benefits of Healthy Food     |
+------------------------------+
4 rows in set (0.00 sec)
```

When you use a where clause, the results are automatically sorted in best first order. If, for example, you do not need a where clause, then assign the relative search ranking to a local variable (here I use score) and add something like order by score desc:

```
mysql> select title, match(contents) against ('health problems organic' ➥
with query expansion) as score from news order by score desc limit 3;
+-------------------------+-----------------+
| title                   | score           |
+-------------------------+-----------------+
| Home Farming News       | 11.355165481567 |
| Benefits of Organic Food | 1.9911414384842 |
```

```
| Benefits of Healthy Food | 1.9911414384842 |
+--------------------------+-----------------+
3 rows in set (0.00 sec)
```

Integrating MySQL Text Search with ActiveRecord

You have already seen my ActiveRecord patches for supporting PostgreSQL full-text search. In this section, I'll integrate the MySQL-specific SQL extensions for full-text search with Active-Record. I will add the same public methods (via monkey patching) to ActiveRecord that I added for PostgreSQL:

```ruby
class ActiveRecord::Base
  def self.text_search_by_column column_name, query
  end
  def self.text_search query
  end
end
```

The first class method `search_by_column` requires you to pass in as an argument the name of the column to search. This is the more efficient API to use because there is no way to tell at runtime using ActiveRecord which columns have full-text indices. The second API, the class method `search`, searches all columns containing text data—this can be very inefficient if you have columns containing text data that do not have a full-text index. The first API is implemented in `mysql-search/mysql-activerecord-simple.rb`. Both APIs are contained in the file `mysql-search/mysql-activerecord.rb`, part of which I show here:

```ruby
class ActiveRecord::Base
  @@searchable_columns = nil
```

The first API that I add to the ActiveRecord base class requires a searchable column name. I build a query string like the examples in the last section, execute the SQL query, and collect the results:

```ruby
def self.text_search_by_column column_name, query
  sql = "select * from " + self.table_name + " where match(" + column_name +
        ") against ('" + query +"' in boolean mode)"
  result = []; ActiveRecord::Base.connection.execute(sql).each {|r| result << r}
  result
end
```

■**Note** The class `Mysql::Result` does not mix-in `Enumerable`, so I could not use the `Enumerable::inject` method in the preceding code.

The next method searches all text columns. The first time that it is called, it sets the class attribute @@searchable_columns using a private method:

```ruby
def self.text_search query
  @@searchable_columns ||= get_searchable_columns
  sql = "select * from " + self.table_name +
          " where match(" + @@searchable_columns +
          ") against ('" + query +"' in boolean mode)"
  result = []; ActiveRecord::Base.connection.execute(sql).each {|r| result << r}
  result
end


private

def self.get_searchable_columns
  ret = []
```

The method `columns` returns `ActiveRecord::ConnectionAdapters::MysqlColumn` objects that I filter, keeping any columns that are of type `text`:

```ruby
  self.columns.each {|f| ret << f.name if f.text?}
  ret.join(',')
  end
end
```

Here is a code snippet using both public methods that I added to the ActiveRecord base class:

```ruby
ActiveRecord::Base.establish_connection(
  :adapter  => :mysql,
  :database => 'test',
  :username => 'root'
)

class News < ActiveRecord::Base
  set_table_name 'news'
end

result = News.text_search_by_column('contents', '+organic')
pp result

result = News.text_search('+organic')
pp result
```

Both `pp result` statements in the preceding code snippet return the same result:

```
[["1",
  "Home Farming News",
  "Both government officials and organic food activists agree that promoting ➡
home and community gardens is a first line of defense during national emergencies"],
 ["3",
  "Benefits of Organic Food",
```

```
"There is now more  evidence that families who eat mostly organic food ➡
may have fewer long term health problems."]]
```

Because both PostgreSQL and MySQL support full-text indexing, your choice of database system is likely to depend on factors other than search. I encourage you to create large sample databases in both MySQL and PostgreSQL and benchmark both systems for the operations that are done most frequently in your applications.

Comparing PostgreSQL and MySQL Text Indexing and Search

I have to admit a small bias toward using PostgreSQL, but on some projects I use MySQL because it is already being used or a customer's server is already set up with it. My extensions to the ActiveRecord base class for MySQL and PostgreSQL full-text search at least provide a common API, but the semantics of the query strings are still different. Remember that even if you use my patches, the form of the query strings differs; for example, compare this first query using the PostgreSQL syntax with the second query using MySQL (in `boolean` mode) full-text search syntax:

```
'bowling | (won & championship)'    # PostgreSQL
'bowling  (+won +championship)'     # MySQL
```

I find the query-string semantics for full-text search in both MySQL and PostgreSQL to be nonintuitive, so I experiment with queries using the `mysql` or `psql` interactive shells.

Wrapup

Search is an important component for most web applications. I hope that you worked along with the examples in this chapter, especially for indexing and search tools that you have not used before. There is no substitute for trying software to see how it fits your requirements.

In the next chapter, you'll learn how to use web scraping and other information-gathering techniques. A common application pattern is collecting data, storing it locally, and providing a search interface.

■ ■ ■

## CHAPTER 10

Using Web Scraping to Create Semantic Relations

Sometimes it is necessary to gather information from web sites that are intended for human readers, not software agents. This process is known as "web scraping."

The first thing you must do when considering web scraping is respect the rights of web-site owners. Different web sites have distinct terms and conditions for using and reusing data hosted on their servers. If your intention is to republish other people's data and information, check with the web-site owners and ask for permission unless their terms of service specifically allow data reuse.

When web-scraping, you must be careful not to submit too many web-page requests in a short time interval. I try never to request more than one page per second from someone else's server, and if possible I reduce this to one page request every five seconds. In addition to simply being considerate of other people's server resources, you face another practical issue: if you make too many service requests or too many within a very short time interval, then web sites might think you're making a denial-of-service attack on a server. It is probable that your IP address will be blocked from making future requests, and you might even face legal implications arising from not being considerate. Speaking of being considerate, consider contacting web-site owners and thanking them for the use of their information. If you are writing technical papers, give credit to web sites providing the data you are using.

I will use two of my own web sites for the examples in this chapter. Both sites contain information about food and recipes. In addition to writing web scrapers for these sites, I'll show you in the final example program how to find common semantic relations between data in these sites and how to generate RDF metadata.

My work involves two activities that require web scraping: basic research in natural language processing and information extraction, and work for customers building customized information-gathering and information-processing software. Twelve years ago, I used to write custom web scrapers in Java. Fortunately, there are now high-level tools for web scraping written in my favorite scripting language (Ruby!) that are far easier to use. For the examples in this chapter, I will use two tools: scRUBYt! and Watir. scRUBYt! was written by Peter Szinek (the technical reviewer for this book), Glenn Gillen, and other contributors. Watir is developed and supported by WatirCraft, LLC.

■**Note** Watir currently works with Ruby 1.8.6 only. scRUBYt! works with both Ruby 1.8.6 and 1.9.1.

If you have never written web-scraping applications before, several aspects of web scraping make it more difficult than you would think. The fundamental problem, and one that has
no good solution, is that the formats of web sites designed for human readers change often.
These changes require that you determine when your web scraping is no longer producing
good results, and that you subsequently modify your web scrapers for new page layouts and
formats. You'll also encounter other problems when web-scraping: many modern web pages
have very complex HTML structures with multiple content areas on a page, CSS detaches
the visual look from the HTML structure, and Ajax modifies HTML content and structure.
The wide-scale adoption of CSS is making this problem simpler because sites are using fewer
HTML tables for formatting purposes.

There are two basic approaches to scraping data from web pages:

- Start at the home page and perform a breadth-first search of all linked pages on the
  same domain, perhaps limited to a specific search depth.

- For commercial sites with left and right sidebars and multiple content areas, you might
  want to hand-craft a web scraper to ignore some information and links on web pages.

I use the first method when web sites have a simple format and most information on a
page is useful. The second approach is best for web sites that contain irrelevant information
that you want to skip over. The examples developed in this chapter use the second approach:
they scrape recipe and ingredient data from my two sites into a database, and then convert the
database into RDF data, defining relations between data found on both sites.

I discuss in the next section the Firefox web-browser plugin Firebug, which you can use to
simplify the job of analyzing HTML structure on web pages.

Using Firebug to Find HTML Elements on Web Pages

If you want to write a web scraper that's customized for the HTML structure of a specific site,
then your first step is to examine the HTML layout for the pages that you need to automatically
process. While you can manually examine HTML source for web pages of interest, there is a
much better way: use the Firefox Firebug extension. I'll assume for these examples that you
have installed both Firefox (http://www.mozilla.com/en-US/firefox/) and the Firebug extension (https://addons.mozilla.org/en-US/firefox/addon/1843).

■**Caution** Firebug might in some cases show you a different HTML structure from what you see when
fetching a page directly. Peter Szinek pointed out to me that Firefox inserts tbody tags in tables. So, be careful: XPaths that you get using Firebug might not always be compatible with the DOM constructed by HTML
parsers like scRUBYt! and Nokogiri.

The first web site that we will scrape in the next section is a fun cooking web site that my wife and I wrote in 2005: http://www.cjskitchen.com. If you access this site in Firefox and then use the Tools ➤ Firebug ➤ Open menu, you will see a display like Figure 10-1. Firebug shows page HTML as a tree display. Whenever you hover the mouse cursor over a tree element in the lower-left part of the browser window, the HTML element on the web page is highlighted. In Figure 10-1 I have the mouse over this HTML element:

```
<a title="Search Recipes" href="?doSearchRecipes">Search Recipes</a>
```

so the Search Recipes tab at the top of the web page is highlighted.

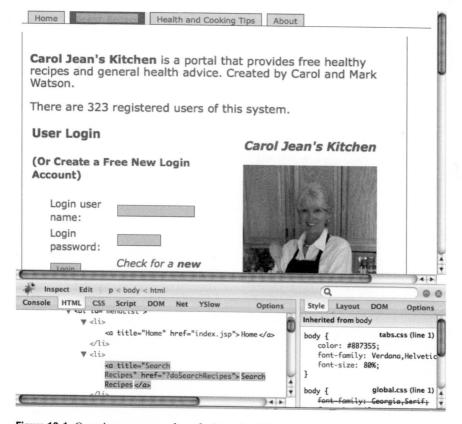

Figure 10-1. *Opening an example web site using Firebug*

Firebug has another feature that you'll find useful: if you right-click any HTML element, you see a popup menu. One of the menu items lets you copy the XPath from the root of the page to the selected HTML element (see Figure 10-2).

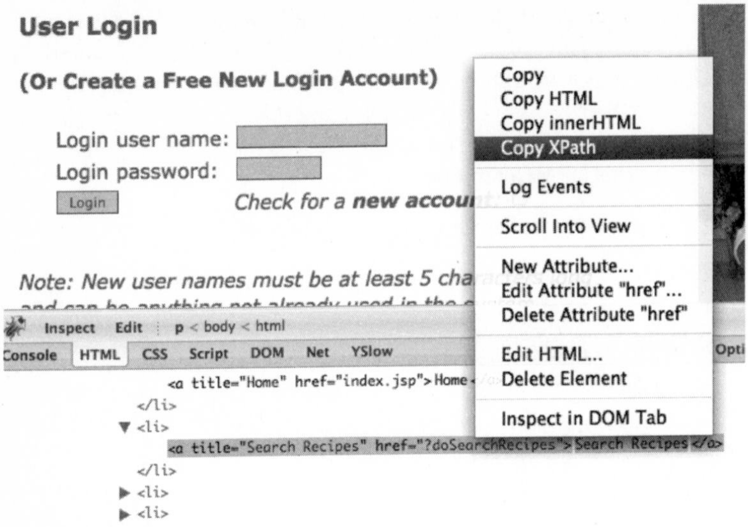

Figure 10-2. *Using Firebug to copy the XPath to any HTML element on a web page*

The XPath to the selected element for the Search Recipes menu item is:

```
/html/body/div/ul/li[2]/a
```

Notice the table of recipes at the bottom of the home web page. The XPath to the recipe on the first row/first column of the HTML table is:

```
/html/body/div[2]/table[2]/tbody/tr/td/a
```

Here's the XPath for a recipe on the first row/third column:

```
/html/body/div[2]/table[2]/tbody/tr/td[3]/a
```

And for one on the third row/fifth column:

```
/html/body/div[2]/table[2]/tbody/tr[3]/td[5]/a
```

Note that there are empty <td/> columns as spacers, so the third visible column is actually the fifth column in the HTML table.

I will introduce you to the scRUBYt! web-scraping library in the next section. We will use the XPaths that we identified in this section to extract recipe and ingredient information.

Using scRUBYt! to Web-Scrape CJsKitchen.com

The scRUBYt! web-scraping library supports a flexible programming model using custom loggers and plugins for building web scrapers. You can get news about new versions at `http://scrubyt.org/blog`. You can also install current versions of the scRUBYt! gem using `gem install scrubyt`. Currently scRUBYt! is at version 0.4.06, but it is being rewritten. I include the current version in the examples directory `src/part3/web-scraping` (download it from the Source Code/Download area of the Apress web site); you should probably initially use this version when trying the examples in this chapter.

Example Use of scRUBYt!

In the last section, you saw the HTML structure for the CJsKitchen.com home page. As an introduction to using scRUBYt!, we will look at finding the recipe descriptions and links from the bottom of the page. Using Firebug, you saw that the XPath to the first recipe description and link was `/html/body/div[2]/table[2]/tbody/tr/td/a`. For the purpose of finding all the recipes at the bottom of the page, you can use the XPath expression `//table/tr/td`.

The scRUBYt! system implements a domain-specific language (DSL) for web scraping. The method `Scrubyt::Extractor.define` processes a code block and returns a value of class `Scrubyt::ScrubytResult`. This result class provides public methods `to_hash` and `to_xml` for exporting data scraped from a page. In the following code snippet, the code block passed to `Scrubyt::Extractor.define` first fetches the web page at `http://cjskitchen.com/` and then defines an extraction data type `recipe` with two fields called `recipe_url` and `recipe_text`. Both of these field names are followed by XPath expressions that are relative to `//table/tr/td`:

```
require 'rubygems' # needed for Ruby 1.8.6
require 'scrubyt'

recipes = Scrubyt::Extractor.define do
  fetch 'http://cjskitchen.com/'

  recipe '//table/tr/td' do
    recipe_url          "//a/@href"
    recipe_text         "//a"
  end
end


recipes.to_hash.each {|hash|
  if hash[:recipe_url]
    puts "#{hash[:recipe_text]} link: http://cjskitchen.com/#{hash[:recipe_url]}"
  end
}
```

If you run this code snippet, the output looks like this:

```
Arroz con Pollo link: http://cjskitchen.com/printpage.jsp?recipe_id=13019321
Asian Chicken Rice link: http://cjskitchen.com/printpage.jsp?recipe_id=1360065
Barbecued Cornish Game Hens link: http://cjskitchen.com/printpage.jsp?recipe_id=➡
```

```
8625327
Cheryl's Marjoram Chicken and Creamy Yogurt Sauce link: http://cjskitchen.com/➥
printpage.jsp?recipe_id=2488285
```

If you use the to_xml method, the generated XML looks like this:

```
<root>
  <recipe>
    <recipe_url>printpage.jsp?recipe_id=13019321</recipe_url>
    <recipe_text>Arroz con Pollo</recipe_text>
  </recipe>
  . . .
</root>
```

Database Schema for Storing Web-Scraped Recipes

While our goal is to publish relations in RDF notation between two cooking sites, we need to store information from the sites for offline processing. Breaking up the processing into multiple steps makes the implementation easier because you can get one step at a time implemented and tested.

The Ruby script create_database_schema.rb, included with this chapter's source code, creates two tables that I will use for intermediate storage. Here I am using the ActiveRecord:: Schema.define utility class to express a database schema, as I did in Chapter 8:

```
require 'rubygems'
require 'activerecord'

ActiveRecord::Base.establish_connection(:adapter => :postgresql, :database => ➥
'test', :username => 'postgres')

ActiveRecord::Schema.define do
  create_table :scraped_recipes do |t|
    t.string :base_url # will be http://knowledgebooks.com or http://cjskitchen.com
    t.integer :recipe_id
    t.string :recipe_name
    t.string :directions
  end

  create_table :scraped_recipe_ingredients do |t|
    t.string :description
    t.integer :recipe_id
  end
end
```

Storing Recipes from CJsKitchen.com in a Local Database

I already showed you a code snippet for using scRUBYt! to fetch the individual recipe-page links from the CJsKitchen.com home page. Now I'll show you the code—interspersed with explanations—that stores recipes from the site in a local database. (You can also find the code in the Ruby script scrubyt_cjskitchen_to_db.rb, included with this chapter's sample files.) Start with the require statements:

```
require 'rubygems' # needed for Ruby 1.8.6
require 'scrubyt'
require 'activerecord'
```

Then open an ActiveRecord connection to PostgreSQL:

```
ActiveRecord::Base.establish_connection(:adapter => :postgresql,
                               :database => 'test', :username => 'postgres')
```

Now generate ActiveRecord classes for two tables in the database, one for recipes (scraped_recipes) and one for ingredients (scraped_recipe_ingredients):

```
class ScrapedRecipe < ActiveRecord::Base
end

class ScrapedRecipeIngredient < ActiveRecord::Base
end
```

The following utility method, recipe_to_db, is called once for each recipe-page link at the bottom of the home page. I used Firebug to determine the XPath patterns for the recipe title, ingredient descriptions, and ingredient amounts:

```
def recipe_to_db(recipe_url)
  recipe = Scrubyt::Extractor.define do
    fetch "http://cjskitchen.com/#{recipe_url}"
    recipe2 '//table/tr' do
      title        "/td[1]/h2"
      description   "/td[1]"
      amount        "/td[2]"
    end
```

The HTML structure for the individual recipe pages has the directions as top-level text in the <body> element, so the following XPath expression gets all of the text on the web page (I will extract the directions from the page text later):

```
    recipe2 '//body' do
      directions "/"
    end
  end
```

The following code gets the recipe ID from the recipe URL and creates a new instance of the class ScrapedRecipe:

```
index = recipe_url.index('=')
puts "recipe_url = #{recipe_url} and index=#{index}"
recipe_id = recipe_url[index+1..-1].to_i
a_recipe = ScrapedRecipe.new(:recipe_id => recipe_id,
                              :base_url => 'http://cjskitchen.com')
```

The next code snippet uses the hash-table representation of a Scrubyt::Extractor instance to define the attributes for the current recipe object, and also to create recipe-ingredient objects and save them to the database:

```
recipe.to_hash.each {|hash|
  if hash[:title]
    a_recipe.recipe_name = hash[:title]
  elsif hash[:directions]
    index = hash[:directions].index('Number of people served by the recipe:')
    a_recipe.directions = hash[:directions][index..-1] if index
  elsif hash[:description]
    ingredient = ScrapedRecipeIngredient.new(
                          :description => hash[:description],
                          :amount => hash[:amount], :recipe_id => recipe_id)
    ingredient.save!
  end
}
```

The last thing that the utility method recipe_to_db needs to do is save the current recipe object to the database:

```
  a_recipe.save!
end
```

The following code (at the bottom of file scrubyt_cjskitchen_to_db.rb) loops over the recipe-page links at the bottom of the home page and calls the utility method recipe_to_db for each recipe URL:

```
recipes = Scrubyt::Extractor.define do
  fetch 'http://cjskitchen.com/'

  recipe '//table/tr/td' do
    recipe_url        "//a/@href"
  end
end

recipes.to_hash.each {|hash|
  recipe_to_db(hash[:recipe_url]) if hash[:recipe_url]
  sleep(1) # wait one second between page fetches
}
```

After you run the script in scrubyt_cjskitchen_to_db.rb, the scraped_recipes table has 31 rows and the scraped_recipe_ingredients table has 248 rows. In the next section, we will use the Watir web-scraping library to scrape the http://cookingspace.com site and add recipes from this site to our database. Later, we will extract relations between recipes on both sites and express these relations in RDF.

Using Watir to Web-Scrape CookingSpace.com

When Watir was first released, it was a Windows-only library because it used OLE to drive Internet Explorer. Fortunately, an extension of the Watir project, FireWatir, now also supports Mac OS and Linux via Firefox 2 and Firefox 3 plugins. You can find detailed, platform-dependent installation instructions at `http://wtr.rubyforge.org/install.html`. On Windows, install using `gem install watir`. On Linux or Mac OS X, install using `gem install firewatir`. You will need to install the Firefox scripting plugin; you can find links for this on the Watir installation web page.

Watir works by scripting Internet Explorer and FireWatir works by scripting Firefox, so it is relatively easy to web-scrape sites requiring logins and the like. Because you can use Fire-Watir on Windows, Linux, and Mac OS X, for the examples in this section I will use FireWatir instead of the Windows-only Watir gem. For most web-scraping work, either scRUBYt! or (Fire)Watir works fine; I am using both of them in this chapter so that you can choose for yourself. The current version of scRUBYt! also works with FireWatir (`http://www.ruby-forum.com/topic/151017`).

Example Use of FireWatir

Our goal is to add recipes and ingredients from CookingSpace.com to the same local database where we added the content from CJsKitchen.com.

The following code snippets are in the file `watir_cookingspace_test.rb`. I start by requiring the FireWatir gem and creating a new browser object. Before executing this script, close all instances of Firefox that are running. If you are using the current version of Firefox, creating an instance of `Watir::Browser` will start Firefox and connect to the scripting plugin:

```
require 'rubygems' # required for Ruby 1.8.6
require "firewatir"

# open a browser
browser = Watir::Browser.new
```

The following statement scripts Firefox to go to my cooking web site `http://cookingspace.com`:

```
browser.goto("http://cookingspace.com")
```

If you look at the HTML source for my site's home page (or use Firebug), you will see this HTML form for searching my site:

```
<form action="/" method="post">
    <input type="text" size="20" name="search_text" />
    <input type="submit" name="search" value="Search"/>
</form>
```

I want to script Firefox to enter search text in the HTML element named search_text and then click the Search button. This will have the same effect as manually typing the search query in the form on Firefox and clicking the Search button:

```
browser.text_field(:name, "search_text").set("salmon")
browser.button(:name, "search").click
```

At this point, I could use the browser method text to access all of the text (with no HTML tags) from the current page that now shows the search results (output is not shown):

```
puts browser.text
```

The following statement prints out all links on the current web page:

```
browser.links.each { |link| puts "\nnext link: #{link.to_s}" }
```

Link objects have accessor methods called text and href that you'll use to get prompt text and the URL for each link. The prompt text is helpful because you can use it to locate a particular link on a web page. In this example, I am accessing the link with the prompt text "Salmon Rice" and then clicking this link to script Firefox to navigate to the linked web page:

```
link = browser.link(:text, "Salmon Rice")
link.click  if link
```

The Firefox browser started by FireWatir should now be showing the detail recipe page for "Salmon Rice." The browser variable contains an instance of class FireWatir::Firefox, which defines almost 100 public methods. The following methods will likely be of use for your projects: body, buttons, check_boxes, divs, html, links, select_lists, tables, and title. Developing with FireWatir should be an interactive experience: run it in an interactive irb session, observe changes to the instance of Firefox as you use the FireWatir APIs, and print out HTML page attributes using the FireWatir::Firefox public methods.

In the next section, you'll store recipe data scraped from CookingSpace.com by using the same database schema and Ruby classes ScrapedRecipe and ScrapedRecipeIngredient that you used to process CJsKitchen.com.

Storing Recipes from CookingSpace.com in a Local Database

CookingSpace.com presents a different web-scraping challenge because it does not offer a list of all recipes. This is a good example because the structure and navigation of every web site is different. Almost every web site for which you'll need to write a custom web scraper will have its own unique problems.

If you wanted to get all recipes from CookingSpace.com, you could choose from three approaches. First, you could use the fact that recipe IDs seem to be assigned consecutively, with some IDs missing because of deleted recipes. The URL pattern for showing recipes is http://cookingspace.com/site/show_recipe/10, where the ending number is the recipe ID in a database. Second, you could use the URL http://cookingspace.com/site/show_random_recipe often enough to cover most of the recipes. Third, you could create a collection of unique words used in recipe names on CJsKitchen.com and use these words one at a time to search for recipes on CookingSpace.com using the FireWatir APIs that you experimented with in the last section.

I am going to use the second option for collecting most of the available recipes on CookingSpace.com. I will keep track of recipes already processed and stop the collection process whenever randomly generating recipes five times in a row does not find an unprocessed recipe.

I did not write CookingSpace.com to be easy to web-scrape (I wrote it for my own use to track the nutrients in my diet). If you look at a fragment of HTML page source, you'll see a few problems that you'll encounter while scraping recipes:

```
<table style="width: 75%">
  <tr>
    <td style="vertical-align: top; width: 40%">
      <strong>Recipe for: Camembert cheese and crackers</strong><br/>
      <p><strong>Ingredients:</strong></p>

          Camembert cheese: 3  tablespoons<br />

          Nabisco Ritz Crackers: 1  serving<br />

      <br/>
      <p><strong>Directions:</strong></p>
      <strong><pre>Spread softened cheese gently on crackers</pre></strong>
    </td>
  </tr>
    ...
</table>
```

The problem is that there is no straightforward way to use XPath to pick off the ingredient descriptions and amounts. So while you can use an XPath expression to get the center content area of the web page, you'll have to use custom string-matching code to parse the text for ingredients. The following code snippet takes advantage of the fact that each recipe page is automatically generated (that is, it doesn't contain human-edited HTML) and looks for regular patterns that will be the same for each recipe display page:

```
def get_recipe_from_page content_html
  in_ingredients = false
  in_directions = false
  ingredients = {}
  name = ''
  directions = ''
  content_html.each_line {|line|
    line = line.strip
    if index = line.index('Recipe for:')
      index2 = line.index('</strong>')
      name = line[index+11...index2].strip
    end
    in_ingredients = false if line == "<br/>" || line.index('Directions:')
    if in_ingredients
```

```
        index = line.index(':')
        ingredients[line[0..index]] =
              line[index+1..-1].gsub('<br />', '').gsub('<br>','').strip if index
      end
      in_ingredients = true if line.index('Ingredients:')
      in_directions = false if line == "</tr>"
      if in_directions
        directions << line.gsub(/<\/?[^>]*>/, "") << ' '        # remove HTML tags
      end
      in_directions = true if line.index('Directions:')

  }
  directions.strip!
  [name, ingredients, directions]
end

pp get_recipe_from_page(s) # s contains the HTML table source shown above
```

The output for this example is:

```
["Camembert cheese and crackers",
 {"Camembert cheese:"=>"3  tablespoons",
  "Nabisco Ritz Crackers:"=>"1  serving"},
 "Spread softened cheese gently on crackers"]
```

The Ruby script watir_cookingspace_to_db.rb finds the recipes on CookingSpace.com, creates instances of ScrapedRecipe and ScrapedRecipeIngredient, and saves them to the same database where you stored recipes and ingredients from CJsKitchen.com. The following code listing contains the entire contents of this script, except for the get_recipe_from_page method that appeared in the previous snippet:

```
require 'rubygems'
require "firewatir"
require 'activerecord'

require 'pp'

ActiveRecord::Base.establish_connection(:adapter => :postgresql, :database => ➥
'test', :username => 'postgres')

class ScrapedRecipe < ActiveRecord::Base
end

class ScrapedRecipeIngredient < ActiveRecord::Base
end

def get_recipe_from_page content_html; ... ; end  # already listed
```

```
ALREADY_PROCESSED_RECIPE_NAMES = [ ]


# open a browser
$browser = Watir::Browser.new

def process_random_recipe # return false if randomly chosen recipe
                                      # has already been processed, otherwise true
  $browser.goto('http://cookingspace.com/site/show_random_recipe')
  element = $browser.element_by_xpath("//tr/td")
  text = element.innerText
  recipe_name, hash, directions =  get_recipe_from_page(element.html)
  if !ALREADY_PROCESSED_RECIPE_NAMES.include?(recipe_name) # checking name
    ALREADY_PROCESSED_RECIPE_NAMES << recipe_name
    recipe_id = $browser.url
    recipe_id = recipe_id[recipe_id.rindex('/')+1..-1].to_i
    a_recipe = ScrapedRecipe.new(:recipe_id => recipe_id,
                                        :base_url => 'http://cookingspace.com',
          :recipe_name => recipe_name, :directions => directions)
    a_recipe.save!
    hash.keys.each {|ingredient_description|
      ingredient = ScrapedRecipeIngredient.new(:description => ➥
ingredient_description,
          :amount => hash[ingredient_description], :recipe_id => recipe_id)
      ingredient.save!
    }
    return true
  end
  false
end

count = 0 # quit if count gets to 10
5000.times {|iter|
  sleep(1) # wait one second between page fetches
  if process_random_recipe
    count = 0
  else
    count += 1
  end
  break if count > 10
  puts "iter = #{iter}  count = #{count}"
}
```

When I populated the local database using this script, it took 52 iterations to find the 32 available recipes.

■Note It is interesting to watch the instance of Firefox that FireWatir controls while you run the script `watir_cookingspace_to_db.rb`.

You have seen that both scRUBYt! and FireWatir make it relatively easy to write custom web scrapers for specific web sites. Now that you have web-scraped recipe and ingredient data from two test sites, I'll devote the remainder of this chapter to developing an example system very much in the spirit of Web 3.0: you'll generate relations, defined using RDF, between recipes found on these two web sites.

Generating RDF Relations

I have spent a fair amount of time in this book discussing how to use RDF. In this section, I am going to show you more examples of a scenario when you might want to use RDF: when you make statements about the relations between data resources on different web sites. All of the example code for this section is in the source file `src/part3/web-scraping/create_rdf_relations.rb`. This example program uses the local database containing recipe data from CJsKitchen.com and CookingSpace.com, and generates both RDF data and Graphviz files to help you visualize the relations that we are expressing in RDF.

Extending the ScrapedRecipe Class

The first step in generating relations between resources on different web sites is deciding which relations are both useful and feasible to calculate. In this example, I calculate two types of relations: similarity between recipe names and similarity between recipe ingredients. Previously, I defined the ActiveRecord subclass `ScrapedRecipe` that contained a single statement: a specification that a scraped recipe contained zero or more scraped ingredients. Now I am going to extend this class with methods that allow an instance of `ScrapedRecipe` to compare itself to another instance and calculate similarities between recipe names and between recipe ingredients.

The method `stem_words` creates lists of word stems for both recipe names and ingredient descriptions. The private method `compare_helper` takes two arrays containing word stems and calculates the similarity of the stems in these two arrays. Both `compare_name_to` and `compare_ingredients_to` use this private utility method:

```ruby
require 'activerecord'
require 'stemmer'

class ScrapedRecipe < ActiveRecord::Base
  has_many :scraped_recipe_ingredients
  attr_accessor :name_word_stems
  attr_accessor :ingredient_word_stems
  @@stop_words = ['and', 'the', 'with', 'frozen', '&amp;', 'salt', 'pepper']
  def stem_words
    @name_word_stems =
      self.recipe_name.downcase.scan(/\w+/).collect {|word|
```

```
                                                           word.stem} - @@stop_words
  @ingredient_word_stems =
    self.scraped_recipe_ingredients.inject([]) {|all, ing|
              all = ing.description.downcase.scan(/\w+/).each {|s| s.stem}} -
      @@stop_words
end
def compare_name_to another_scraped_recipe
  compare_helper(self.name_word_stems, another_scraped_recipe.name_word_stems)
end
def compare_ingredients_to another_scraped_recipe
  compare_helper(self.ingredient_word_stems,
                          another_scraped_recipe.ingredient_word_stems)
end
private
def compare_helper list1, list2
```

The following statement calculates a normalized comparison value for lists of arbitrary sizes:

```
  2.0 * (list1 & list2).length / (list1.length + list2.length + 0.01)
end
end
```

The expression `list1 & list2` produces an array that contains elements found in both `list1` and `list2`. The denominator normalizes by list sizes.

Figure 10-3 shows a Graphviz visualization of the recipe-name and recipe-ingredient similarities. It shows only links with larger similarity values; reducing the link-strength threshold produces graphs with more nodes and links. The code to create these visualizations uses the Graphviz gem, which you must install with `gem install graphviz`.

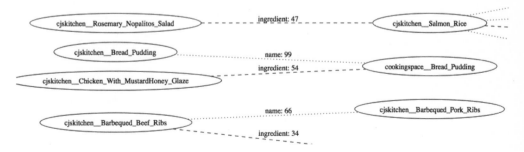

Figure 10-3. *Graphviz view of recipe similarities*

Graphviz Visualization for Relations Between Recipes

You already used Graphviz in Chapters 4 and 7, so I will use it here without much explanation. The visualizations are useful for showing people the relations you can determine from extracted data. Whereas human readers will find these visualizations helpful, software agents will benefit more from the similar RDF graphs that I'll generate later from the same data and relations.

The following code snippet uses the "compare to" methods in the class `ScrapedRecipe` that I developed in the last section to generate Graphviz input files:

```ruby
require 'graphviz'; require 'activerecord'

ActiveRecord::Base.establish_connection(:adapter => :postgresql,
                                         :database => 'test', :username => 'postgres')
$recipes = ScrapedRecipe.find(:all)
$ingredients = ScrapedRecipeIngredient.find(:all)

# stem words in all recipes:
$recipes.each {|recipe| recipe.stem_words}

# generate Graphviz dot file output:

g = GraphViz::new("G", "output" => 'dot')
g["rankdir"] = "LR"
g.node["shape"] = "ellipse"

# convert a scraped recipe to a display name (web site name followed by recipe name)
# Spaces and other special characters should not be in Graphviz node names:

def get_display_name a_recipe
  rn = $recipes[i].recipe_name.strip.gsub(' ','_').gsub("'s", "").gsub("&amp;",
          "and").gsub('(','').gsub(')','').gsub(',','').gsub('-','')
  if $recipes[i].base_url.index('cjskitchen')
    rn = 'cjskitchen__' + rn
  else
    rn = 'cookingspace__' + rn
  end
  rn
end

# perform a double loop over all scraped recipes, comparing each recipe
# to all other recipes (warning: scales as O(n^2), where n==number of recipes):

num = $recipes.length
num.times {|i|
  num.times {|j|
    if j > i
      rn1 = get_display_name($recipes[i])
      rn2 = get_display_name($recipes[j])
      name_similarity = $recipes[i].compare_name_to($recipes[j])
      ingredients_similarity = $recipes[i].compare_ingredients_to($recipes[j])
      if name_similarity > 0.4
        # add Graphviz nodes and links:
```

```
        g.add_node(rn1)
        g.add_node(rn2)
        g.add_edge(rn1, rn2, :arrowhead => "none", :style => "dotted",
                            :label => "name: #{(100.0*name_similarity).to_i}")
      end
      if ingredients_similarity > 0.3
        # add Graphviz nodes and links:
        g.add_node(rn1)
        g.add_node(rn2)
        g.add_edge(rn1,rn2, :arrowhead => "none", :style => "dashed",
                          :label => "ingredient: #{(100.0*➥
                          ingredients_similarity).to_i}")
      end
    end
  }
}
g.output(:file => "recipes.dot")
```

In this example code snippet, I am generating Graphviz nodes if the recipe-name similarity is greater than 0.4 or the ingredient similarity between two scraped recipes is greater than 0.3. I created Figure 10-4 by reducing these cutoff values to 0.001. You can experiment with these cutoff values to change the number of nodes and links that are generated.

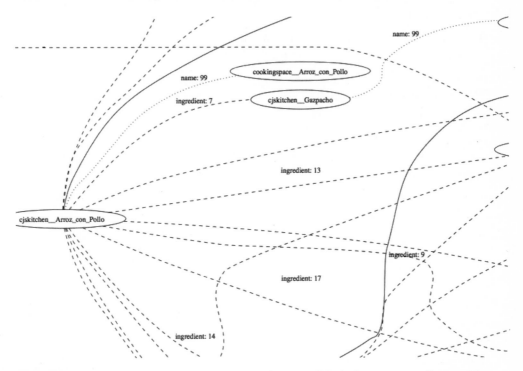

Figure 10-4. *Graphviz view of recipe similarities, showing all links for a very small area of the graph display*

RDFS Modeling of Relations Between Recipes

I find the Graphviz visualizations to be very useful for seeing calculated relations, as you just saw in the recipe-similarity comparisons. I am now going to generate RDF to specify these relations. The RDF notation will be much richer than the visualizations in Figures 10-3 and 10-4 because I collapsed the web-site source and the recipe name into a single node name. I also collapsed both the relation type and the relation strength into link names. In RDF, this data will be stated explicitly. We will start with comparing two nodes with a single link, as shown in Figure 10-5 with the equivalent RDF notation.

Figure 10-5. *Graphviz view of recipe-name similarity between two recipes*

When you model these relations in RDF, your first decision is whether you want to pre-serve the numeric rating of the similarity links. Modeling numeric rating is fairly simple: you could create unique resources for each similarity link and make statements regarding the link type and numeric value. While SPARQL allows you to query on numeric ranges, there are at least two good reasons for excluding numeric values from similarities:

- Including numeric values will hinder scalability. Querying/inferencing over very large data sets will be much more efficient when you match on property (relation) type only, rather than on both property type and the strength of the relation.

- It makes sense to simply make statements like "recipe A has a similar name to recipe B" without concern for *how similar* they are.

Let's look at this issue from the point of view of a developer writing an application using RDF. You want to be able to find information that is similar to a known resource on the Web. Ideally, you can find RDF data making statements about this known resource that compare it using well-defined properties to other resources on the Web. We will revisit this concept, known as *Linked Data*, in Chapter 11.

In RDF N3 notation, you can express Figure 10-5 as RDF data in the manually edited source file sample.n3:

```
@prefix kb: <http://knowledgebooks.com/test#> .
@prefix rdf: <http://www.w3.org/1999/02/22-rdf-syntax-ns#> .
@prefix rdfs: <http://www.w3.org/2000/01/rdf-schema#> .

kb:Recipe a rdfs:Class;
        rdfs:label "Recipe".

kb:similarRecipeName a rdf:Property ;
                rdfs:domain kb:Recipe;
                rdfs:range kb:Recipe .
```

```
kb:similarRecipeIngredients a rdf:Property ;
                            rdfs:domain kb:Recipe;
                            rdfs:range kb:Recipe .

kb:recipeName a rdf:Property;
              rdfs:domain kb:Recipe;
              rdfs:range <http://www.w3.org/2001/XMLSchema#stringstring> .

<http://cjskitchen.com/printpage.jsp?recipe_id=8625327> a kb:Recipe .
<http://cjskitchen.com/printpage.jsp?recipe_id=8625327>
   kb:recipeName
   "Barbecued Cornish Game Hens" .

<http://cookingspace.com/site/show_recipe/3> a kb:Recipe .
<http://cookingspace.com/site/show_recipe/3>
   kb:recipeName
   "Barbecued Cornish Game Hens" .

<http://cjskitchen.com/printpage.jsp?recipe_id=8625327>
   kb:similarRecipeName
   <http://cookingspace.com/site/show_recipe/3> .

<http://cookingspace.com/site/show_recipe/3>
   kb:similarRecipeName
   <http://cjskitchen.com/printpage.jsp?recipe_id=8625327> .
```

The last two statements express that the kb:similarRecipeName property is symmetric. I could have used the OWL property owl:SymmetricProperty to reduce the redundancy of making two statements, but I wanted to use just RDF/RDFS so I used two separate statements to express a symmetric link.

It is useful to look at some example SPARQL queries using the RDF in the sample.n3 file:

```
PREFIX rdfs:<http://www.w3.org/2000/01/rdf-schema#>
PREFIX kb:<http://knowledgebooks.com/test#>
PREFIX rdf:<http://www.w3.org/1999/02/22-rdf-syntax-ns#>

select distinct ?s ?p ?o where {
   ?s a kb:Recipe .
   ?s  ?p ?o .
}
```

This example finds 10 triples:

```
<http://cjskitchen.com/printpage.jsp?recipe_id=8625327>
   rdf:type
   rdfs:Resource .
```

```
<http://cjskitchen.com/printpage.jsp?recipe_id=8625327>
  rdf:type
  kb:Recipe .

<http://cjskitchen.com/printpage.jsp?recipe_id=8625327>
  kb:similarRecipeName
  <http://cookingspace.com/site/show_recipe/3> .

<http://cjskitchen.com/printpage.jsp?recipe_id=8625327>
  kb:recipeName
  "Barbecued Cornish Game Hens" .

<http://cjskitchen.com/printpage.jsp?recipe_id=8625327>
  <http://www.openrdf.org/schema/sesame#directType>
  kb:Recipe .

<http://cookingspace.com/site/show_recipe/3>
  rdf:type
  rdfs:Resource .

<http://cookingspace.com/site/show_recipe/3>
  rdf:type
  kb:Recipe .

<http://cookingspace.com/site/show_recipe/3>
  kb:similarRecipeName
  <http://cjskitchen.com/printpage.jsp?recipe_id=8625327> .

<http://cookingspace.com/site/show_recipe/3>
  kb:recipeName
  "Barbecued Cornish Game Hens" .

<http://cookingspace.com/site/show_recipe/3>
  <http://www.openrdf.org/schema/sesame#directType>
  kb:Recipe .
```

The sample.n3 file was manually edited. In the next section, you'll generate a similar but much larger file from the data in our local recipes and ingredients database.

Automatically Generating RDF Relations Between Recipes

Having the sample.n3 file that I manually created makes it easy to add code to the example script create_rdf_relations.rb to write out generated RDF to a local file. Manually generating a small sample file and testing it (I used the Sesame Workbench for this) will save you a lot of time because errors are easy to correct with a small hand-edited file. I am not going to list the entire create_rdf_relations.rb file here. Instead, the following snippets show only the code that I added to generate the RDF data. (You can get the full script by downloading the source code for this chapter from the Apress web site.) The following code writes out the RDFS type and property definitions from a static string:

```
fout = File.open('recipes.rdf', 'w')
fout.puts('@prefix kb: <http://knowledgebooks.com/test#> .
@prefix rdf: <http://www.w3.org/1999/02/22-rdf-syntax-ns#> .
@prefix rdfs: <http://www.w3.org/2000/01/rdf-schema#> .

kb:Recipe a rdfs:Class;
    rdfs:label "Recipe".

kb:similarRecipeName a rdf:Property ;
    rdfs:domain kb:Recipe;
    rdfs:range kb:Recipe .

kb:similarRecipeIngredients a rdf:Property ;
    rdfs:domain kb:Recipe;
    rdfs:range kb:Recipe .

kb:recipeName a rdf:Property;
    rdfs:domain kb:Recipe;
    rdfs:range <http://www.w3.org/2001/XMLSchema#stringstring> .

')
```

The following method, get_recipe_full_uri, converts a ScrapedRecipe object that has been fetched from the local database to a properly formatted RDF resource URI such as <http://cookingspace.com/site/show_recipe/2> or <http://cjskitchen.com/printpage.jsp?recipe_id=3435721>:

```
 def get_recipe_full_uri a_recipe
  return "<#{a_recipe.base_url}/printpage.jsp?recipe_id=#{a_recipe.recipe_id}> "➡
         if a_recipe.base_url.include?('cjskitchen')
  "<#{a_recipe.base_url}/site/show_recipe/#{a_recipe.recipe_id}> "
end

num = $recipes.length
num.times {|i|
  fout.puts(get_recipe_full_uri($recipes[i]))
  fout.puts("  a kb:Recipe ;")
  # add the recipe name:
  fout.puts("  kb:recipeName \"#{$recipes[i].recipe_name}\" .")
  num.times {|j|
    if j > i
      if name_similarity > 0.4
        # write on N3 RDF:
        fout.puts("#{get_recipe_full_uri($recipes[i])} kb:similarRecipeName  ➡
                    #{get_recipe_full_uri($recipes[j])} .")
        fout.puts("#{get_recipe_full_uri($recipes[j])} kb:similarRecipeName  ➡
```

```
                          #{get_recipe_full_uri($recipes[i])} .")
        end
      if ingredients_similarity > 0.3
        # write on N3 RDF:
        fout.puts("#{get_recipe_full_uri($recipes[i])} ➥
                      kb:similarRecipeIngredients ➥
                      #{get_recipe_full_uri($recipes[j])} .")
        fout.puts("#{get_recipe_full_uri($recipes[j])} ➥
                      kb:similarRecipeIngredients ➥
                      #{get_recipe_full_uri($recipes[i])} .")
      end
    end
  }
}
fout.close
```

This code is similar to the code you used to generate Graphviz visualizations from the same database. Check out src/part3/web-scraping/recipes.dot to see a sample file produced with the previous code snippet. A few lines of output look like this:

```
digraph G {
    graph [rankdir=LR];
    node [label="\N", shape=ellipse];
    cjskitchen__Arroz_con_Pollo [pos="259,18", width="3.50", height="0.50"];
    cjskitchen__Arroz_con_Pollo ->
                  cookingspace__Arroz_con_Pollo [label="name: 99", arrowhead=none];
```

■**Note** The create_rdf_relations.rb file contains the code for producing both Graphviz and RDF output.

Let's try a sample query to find recipes with similar ingredients to a known recipe resource:

```
PREFIX rdfs:<http://www.w3.org/2000/01/rdf-schema#>
PREFIX kb:<http://knowledgebooks.com/test#>
PREFIX rdf:<http://www.w3.org/1999/02/22-rdf-syntax-ns#>

select distinct ?o where {
   <http://cjskitchen.com/printpage.jsp?recipe_id=1804194> ➥
kb:similarRecipeIngredients ?o
}
```

This query finds a single similar recipe:

```
<http://cjskitchen.com/printpage.jsp?recipe_id=5078181>
```

If you want to find the name of the matching recipe, you could use this query:

```
PREFIX rdfs:<http://www.w3.org/2000/01/rdf-schema#>
PREFIX kb:<http://knowledgebooks.com/test#>
PREFIX rdf:<http://www.w3.org/1999/02/22-rdf-syntax-ns#>

select distinct ?recipe_name where {
   <http://cjskitchen.com/printpage.jsp?recipe_id=5078181>
      kb:recipeName
      ?recipe_name .
}
```

The query result is: "Mark's Baked Lamb Shanks."

Publishing the Data for Recipe Relations as RDF Data

Publishing data and providing web services is the subject of Part 4 of this book. But as you saw in Chapter 5, publishing RDF data using Sesame web services is simple, and mentioning it again here is worthwhile. When developing this sample program, I used the Sesame Workbench to test both my hand-edited N3 RDF file `sample.n3` and the generated file `recipes.n3`. I created a repository with an ID of 101 (an arbitrary number that could also have been a more meaningful string constant), so the recipe similarity RDF data is available if you use the root URL `http://localhost:8080/openrdf-sesame/repositories/101?query=` and append a CGI-encoded SPARQL query.

To publish this generated RDF data on the Web, you would run Sesame web services on a public server and publish the root URL for SPARQL queries and the web-form interface for interactively performing SPARQL queries. If you do not want to allow remote users to modify your RDF data stores, then configure Sesame in "federated mode" and set the read-only option. An easier solution is using the SPARQL endpoint server that I developed in Chapter 7 using Redland and Sinatra. Another good alternative, which I will show you in Chapter 11, is using the D2R system to provide a SPARQL interface for existing relational databases.

Comparing the Use of RDF and Relational Databases

I am almost certain that you are wondering if it is worth the effort to use RDF/RDFS data stores instead of relational database systems. If you are writing a local application, even one that collects data from the Web, then certainly using a relational database is a good approach.

However, for Web 3.0 (Semantic Web) applications, applications are not localized. If you create a local application using a relational database, then your options for providing remote access to your data are limited to:

- Dumping your data into a portable format and making it available

- Providing a custom web-service interface (and documentation) for accessing the information that you want to share

- Providing a web interface and allowing remote users to write web scrapers to extract what they need from your web pages

I believe that using RDF is a better solution for most systems that deal with information resources on the Web. You will see some interesting and useful Linked Data projects in the next chapter. The original Web gave us linked web pages that we could navigate to and read. Web 2.0 mostly dealt with user interface improvements and allowed the use of user input and data to extend the utility of web sites. Web 3.0 is the future, enabling you to link data in new ways and express information resources in ways that are useful for both human readers and software consumers of data.

Wrapup

You learned how to use two popular web-scraping libraries in this chapter. More important, I showed you one useful application for web scraping: calculating metadata from scraped content. This metadata makes the original information more useful because it allows you to find new, similar information and determine the relation between the original and new information.

I believe that in the future, there will be a large market for semantic metadata that describes information on the Web, both for free use and for fee-based use. While I do show you in the next chapter how to wrap relational databases as SPARQL endpoints, I encourage you to view RDF as a first choice in publishing Linked Data in your Web 3.0 applications.

CHAPTER 11

■■■

Taking Advantage of Linked Data

Linked Data refers to links between data sources, as well as the practice of connecting data on the Web. I have already discussed one part of Linked Data: using RDF to make statements about properties or data resources on the Web and to make statements about the relations between resources. I think that Linked Data will be a key Web 3.0 technology as more Linked Data resources become available. Resources you'll find useful today include DBpedia, Freebase, and Open Calais, and in this chapter you'll see examples of using Linked Data from all three of these. You will also see that it is easy to publish content in relational databases as Linked Data using the D2R system. (I'll use D2R here and in Chapter 15.) Using Linked Data can expand the scope of your Web 3.0 applications to use a wider variety of information sources than you can create with your own resources.

Tim Berners-Lee literally invented the World Wide Web, and with James Hendler and Ora Lassila he wrote an influential *Scientific American* article in 2001 popularizing the ideas behind the Semantic Web (http://www.sciam.com/article.cfm?id=the-semantic-web). More recently, Berners-Lee introduced the idea of Linked Data, using four "rules" (http://www.w3.org/DesignIssues/LinkedData.html):

- Use URIs as names for things and concepts.

- Use HTTP URIs so that people can look up those names; you can "dereference" URIs by retrieving data using an HTTP GET on the URI.

- When someone retrieves data referenced by a URI, provide useful information. This can be HTML for human readers or RDF for software agents, depending on the HTTP GET request headers.

- Include links to other URIs so that both human readers and software agents can discover more things related to a URI.

■**Tip** In Appendix B, I will cover details for publishing Linked Data as URIs that can be "dereferenced." Dereferencing a URI involves fetching the URI contents. If a web browser accesses a URI, then you want to return neatly formatted HTML for human readers. If a URI GET request comes from a client application, then you want to return RDF data with the HTTP response headers indicating the type of data (for example, application/rdf+xml or text/rdf+n3).

What does it mean to use URIs as names for things and concepts? As an example, if I want to name myself, I might use the URI http://markwatson.com to reference my entire web site—or even better, I could reference my site's RDF metadata using http://www.markwatson.com/index.rdf. If someone dereferences my site metadata, he will get back a static RDF document that uses three standard ontologies for making RDF statements about me.

Now we will look at additional techniques for linking resources on the Web and examples of Linked Data projects. I begin this chapter by continuing the example from the last chapter. Specifically, I'll show you how to use the D2R system to wrap a local database as a SPARQL endpoint. I'll then describe the three public Linked Data sources that I mentioned earlier: DBpedia, Freebase, and Open Calais. You'll learn more publishing techniques in Part 4.

Producing Linked Data Using D2R

The last example in Chapter 10 showed you how to use a local database to store data scraped from two web sites, and you used this local database to generate RDF for publishing relations between the two sites. I will show you a nice alternative in this section: using the D2R system (http://www4.wiwiss.fu-berlin.de/bizer/d2r-server/) to provide a SPARQL interface for existing relational databases. As an example, I'll use the PostgreSQL database test and the two tables scraped_recipes and scraped_recipe_ingredients that I created in Chapter 10. I will publish the data in these two tables via a SPARQL endpoint.

To follow along with this example, download the D2R distribution. If you have Java installed on your system, there are no other required installation steps because the D2R distribution is otherwise self-contained. The following instructions to create a mapping file and start the server apply specifically to the Chapter 10 example database. You only need to create the mapping file once unless the database schema changes:

```
cd d2r-server-0.6
generate-mapping -o mapping.n3 -d org.postgresql.Driver    ➥
                          -u postgres jdbc:postgresql://localhost/test
d2r-server mapping.n3
```

You now have a local SPARQL endpoint service and also a web interface available at http://localhost:2020/. An example D2R web page is shown in Figure 11-1.

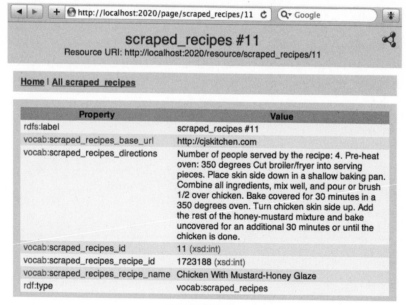

Figure 11-1. *Using the D2R web interface*

The D2R server also has an interactive SPARQL query window similar to that in the Sesame web interface. The mapping file that D2R creates for you defines N3-prefix shortcuts for the namespaces that it establishes for your database. It also uses existing standard namespaces:

```
PREFIX xsd: <http://www.w3.org/2001/XMLSchema#>
PREFIX rdf: <http://www.w3.org/1999/02/22-rdf-syntax-ns#>
PREFIX rdfs: <http://www.w3.org/2000/01/rdf-schema#>
PREFIX owl: <http://www.w3.org/2002/07/owl#>

PREFIX db: <http://localhost:2020/resource/>
PREFIX vocab: <http://localhost:2020/vocab/resource/>
```

The prefix db: is used for URIs defined from a table name and a primary key. Each URI with the prefix db:: represents a single row in the database. Here are two examples:

```
db:scraped_recipes/1
db:scraped_recipe_ingredients/261
```

The db: prefix always appears in a URI as the subject of an RDF triple. The prefix vocab:, on the other hand, defines properties based on combining table and column names. Each URI with the vocab: prefix defines a unique column in the database, as in these examples:

```
vocab:scraped_recipes_base_url
vocab:scraped_recipes_id
vocab:scraped_recipes_directions
vocab:scraped_recipes_recipe_name
vocab:scraped_recipes_recipe_id

vocab:scraped_recipe_ingredients_description
vocab:scraped_recipe_ingredients_scraped_recipe_id
vocab:scraped_recipe_ingredients_amount
vocab:scraped_recipe_ingredients_id
```

The vocab: prefix is also used to define classes:

```
vocab:scraped_recipes
vocab:scraped_recipe_ingredients
```

By manually editing the mapping file, you can change the property and class names used to create virtual RDF triples from data in a database. As a developer, I find the automatically generated names easy to understand because I am already familiar with the database schema. If you're publishing a SPARQL endpoint for external users, you might want to follow the instructions on the D2R web site for modifying your mapping file.

You can use the SPARQL-query web form at http://localhost:2020/snorql to get the "virtual" RDF triples created from my recipe database. The default query prints all triples with a limit of 10 results; I changed this limit to 3,000 to see all "virtual" triples representing my database. Figure 11-2 shows the SPARQL-query web page, executing the following query that shows results for all recipe names and all ingredient descriptions and amounts in all recipes:

```
SELECT DISTINCT ?recipe_name ?ingredient_name ?ingredient_amount WHERE {
    ?recipe_row vocab:scraped_recipes_recipe_name ?recipe_name .
    ?recipe_row vocab:scraped_recipes_id ?recipe_id .
    ?ingredient_row vocab:scraped_recipe_ingredients_scraped_recipe_id ?recipe_id .
    ?ingredient_row vocab:scraped_recipe_ingredients_description ?ingredient_name .
    ?ingredient_row vocab:scraped_recipe_ingredients_amount ?ingredient_amount .
}
LIMIT 3000
```

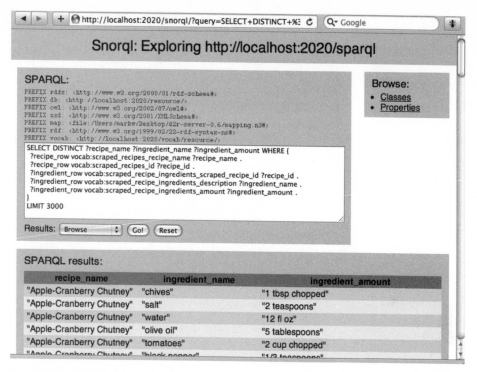

Figure 11-2. *D2R's Snorql SPARQL browsing interface*

This example SPARQL endpoint does not have relations information for similar recipes because the example program in Chapter 10 wrote these relations as external RDF data. You could modify the example in Chapter 10 to write the recipe similarity relations to a third table and then re-create the D2R mapping file.

Now I'll show you how to use a Ruby SPARQL client with a D2R SPARQL endpoint. The example files for the rest of this section are in the directory src/part3/d2r_sparql_client (download the source code for this chapter from the Apress web site at http://www.apress. com). I am reusing the pure Ruby SPARQL client from Chapter 4. The following code snippet, which you can find in test_query.rb, uses the query shown in Figure 11-2:

```
require "rubygems"
require "sparql_client"

qs="
PREFIX rdfs: <http://www.w3.org/2000/01/rdf-schema#>
PREFIX db: <http://localhost:2020/resource/>
PREFIX owl: <http://www.w3.org/2002/07/owl#>
PREFIX xsd: <http://www.w3.org/2001/XMLSchema#>
PREFIX map: <file:/Users/markw/Desktop/d2r-server-0.6/mapping.n3#>
PREFIX rdf: <http://www.w3.org/1999/02/22-rdf-syntax-ns#>
PREFIX vocab: <http://localhost:2020/vocab/resource/>
```

```
SELECT DISTINCT ?recipe_name ?ingredient_name ?ingredient_amount WHERE {
  ?recipe_row vocab:scraped_recipes_recipe_name ?recipe_name .
  ?recipe_row vocab:scraped_recipes_id ?recipe_id .
  ?ingredient_row vocab:scraped_recipe_ingredients_scraped_recipe_id ?recipe_id .
  ?ingredient_row vocab:scraped_recipe_ingredients_description ?ingredient_name .
  ?ingredient_row vocab:scraped_recipe_ingredients_amount ?ingredient_amount .
}
LIMIT 1
"
endpoint="http://localhost:2020/sparql"
sparql = SPARQL::SPARQLWrapper.new(endpoint)
sparql.setQuery(qs)
ret = sparql.query()
puts ret.response
```

I only asked for one triple to be returned, so the XML query results are short enough to list:

```xml
<?xml version="1.0"?>
<sparql xmlns="http://www.w3.org/2005/sparql-results#">
  <head>
    <variable name="recipe_name"/>
    <variable name="ingredient_name"/>
    <variable name="ingredient_amount"/>
  </head>
  <results>
    <result>
      <binding name="recipe_name">
        <literal>Apple-Cranberry Chutney</literal>
      </binding>
      <binding name="ingredient_name">
        <literal>chives</literal>
      </binding>
      <binding name="ingredient_amount">
        <literal>1  tbsp chopped</literal>
      </binding>
    </result>
  </results>
</sparql>
```

D2R is a great tool for publishing Linked Data. Now that you have finished extending the example started in Chapter 10, I will show you the three aforementioned Linked Data sources that you can use in your applications. In Part 4, I'll continue covering techniques for publishing data.

Using Linked Data Sources

In fact, you have already used Linked Data. In Chapter 4 you used SPARQL endpoints for the University of Toronto's Linked Movie Database and the Free University of Berlin's World Fact Book, and you also used the Open Calais system. Now you'll examine two more public Linked Data sources: DBpedia (structured data from Wikipedia) and Freebase (web-service access to structured data). You'll also use Open Calais to generate common URIs for named entities across multiple information sources. In Chapter 4 I used just the comment headers returned from Open Calais web-service calls, but in this chapter I'll use the returned RDF payloads.

DBpedia

The purpose of the DBpedia (`http://dbpedia.org`) project is to extract structured information from Wikipedia and publish it. I find that while DBpedia is an interesting idea, the range of information types might be too broad and the information itself might be too thin, except for a few areas. DBpedia is created and maintained as a community effort, and the topics covered reflect the interests of the community. That said, it is well worth the time to explore DBpedia to see how a large RDF data set is created from many different existing ontologies. These ontologies are represented by different RDF namespaces.

■**Note** The OWL, RDFS, RDF, and XML Schema vocabularies are W3C standards and are represented in separate namespaces. You can find more information about the Friend of a Friend (FOAF) namespace at `http://www.foaf-project.org`. I've been using the Dublin Core Metadata Initiative (DC) vocabulary for describing published data; you can find more information at `http://dublincore.org`. The Simple Knowledge Organization System (SKOS) defines vocabularies for a wide range of knowledge; more information is available at `http://www.w3.org/2004/02/skos`.

The DBpedia public SPARQL endpoint supports the same Snorql web interface for interactive SPARQL queries that you used with D2R: `http://dbpedia.org/snorql`. For the example SPARQL queries in this section, I'll assume that you are trying them yourself using either the Snorql endpoint or the Ruby client code in `src/part3/dbpedia_client`. All queries in this section are assumed to use the following namespace prefixes:

```
PREFIX owl: <http://www.w3.org/2002/07/owl#>
PREFIX xsd: <http://www.w3.org/2001/XMLSchema#>
PREFIX rdfs: <http://www.w3.org/2000/01/rdf-schema#>
PREFIX rdf: <http://www.w3.org/1999/02/22-rdf-syntax-ns#>
PREFIX foaf: <http://xmlns.com/foaf/0.1/>
PREFIX dc: <http://purl.org/dc/elements/1.1/>
PREFIX : <http://dbpedia.org/resource/>
PREFIX dbpedia2: <http://dbpedia.org/property/>
PREFIX dbpedia: <http://dbpedia.org/>
PREFIX skos: <http://www.w3.org/2004/02/skos/core#>
```

Resources in DBpedia have the following base URI: `http://dbpedia.org/resource/`. According to the DBpedia web site, it features URI resources for more than 213,000 people, 328,000 places, 57,000 music albums, 36,000 films, and 20,000 companies. You can try to reference commonly known names by substituting underscore characters for space characters. For example, I just tried six resource names and DBpedia had resource definitions for all of them:

```
http://dbpedia.org/resource/Barack_Obama
http://dbpedia.org/resource/Bill_Clinton
http://dbpedia.org/resource/Paris
http://dbpedia.org/resource/Software
http://dbpedia.org/resource/Cheese
http://dbpedia.org/resource/Health_Insurance
```

Because DBpedia is a Linked Data source, you can dereference any URI to get detailed data on the URI. You can interactively explore the DBpedia data set; in the meantime, I will get you started with example queries. The following SPARQL query finds all people in the DBpedia data set who were born in San Francisco:

```
SELECT ?person WHERE {
    ?person dbpedia2:birthPlace <http://dbpedia.org/resource/San_Francisco> .
}
```

This next query finds all people in the DBPedia data set who were born in Maine and who died in California (three were found):

```
SELECT ?person WHERE {
    ?person dbpedia2:birthPlace <http://dbpedia.org/resource/Maine> .
    ?person dbpedia2:deathPlace <http://dbpedia.org/resource/California> .
}
```

And this query lists the occupations of all 38 people who were born in Maine:

```
SELECT ?person ?occupation WHERE {
    ?person dbpedia2:occupation ?occupation .
    ?person dbpedia2:birthPlace <http://dbpedia.org/resource/Maine> .
}
```

The results from the last example query are shown in Figure 11-3.

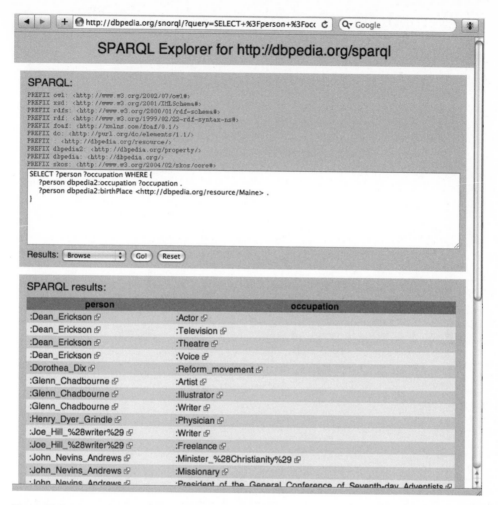

Figure 11-3. *Using the DBpedia Snorql web interface*

Again, DBpedia is a Linked Data source, so exploring the URIs returned from search queries yields additional information. If you click a person result link for "John Nevins Andrews" on the screen shown in Figure 11-3, you'll discover that this link executes another SPARQL query:

```
SELECT ?property ?hasValue ?isValueOf
WHERE {
  { <http://dbpedia.org/resource/John_Nevins_Andrews> ?property ?hasValue }
  UNION
  { ?isValueOf ?property <http://dbpedia.org/resource/John_Nevins_Andrews> }
}
```

The results of this new SPARQL query are shown in Figure 11-4.

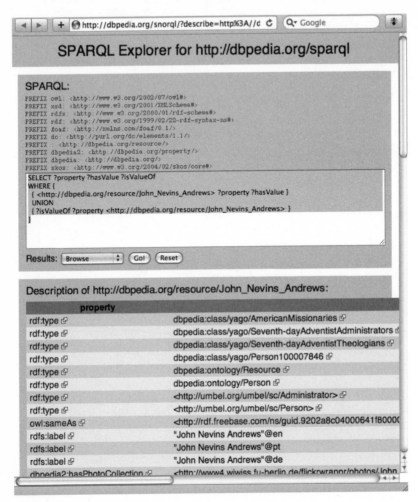

Figure 11-4. *Using the DBpedia Snorql web interface to explore a result URI seen in Figure 11-3*

The first thing that I usually do when working with an unfamiliar SPARQL endpoint is to look for predicates used in the repository. Assuming the use of the namespace abbreviations shown in Figures 11-3 and 11-4, the following query returns the first 1,000 distinct predicates:

```
SELECT distinct  ?pred WHERE {
    ?a ?pred ?o .
}
LIMIT 1000
```

Some of the results for this query are:

```
dbpedia2:genre
dbpedia2:label
dbpedia2:artist
dbpedia2:cover
dbpedia2:released
dbpedia2:reviews
dbpedia2:lastAlbum
dbpedia2:nextAlbum
dbpedia2:recorded
dbpedia2:name
```

You might start exploring this data set by finding information on artists and records:

```
SELECT distinct ?s ?o WHERE {
    ?s dbpedia2:artist ?o  .
}
LIMIT 40
```

Some of the output for this query is:

```
:Silver_Scooter_/_Cursive      " Cursive and Silver Scooter"@en
:Vois_comme_c%27est_beau       " Céline Dion and Claudette Dion"@en
:Les_yeux_de_la_faim           " Céline Dion and various artists"@en
:I_Miss_You_%28DMX_song%29     " DMX featuring Faith Evans"@en
```

■ **Tip** You can download all of the RDF triple files for the DBpedia data sets from the download page at http://wiki.dbpedia.org and load them into an RDF data store such as Sesame. This allows you to experiment with DBpedia locally, which offers convenience as well as protection from DBpedia server downtimes.

I encourage you to read the documentation and look at the examples on the http://wiki.dbpedia.org site. Because the material in DBpedia is contributed by the users who enter structured web pages into Wikipedia, topics are covered in different depths. This variable level of detail might make DBpedia unsuitable in the short term for a knowledge source of real-world data. We will look at another community-driven knowledge source in the next section.

Freebase

Freebase (http://www.freebase.com) is meant to serve as a database containing information about the world that can be accessed through both a web interface and programming APIs. Freebase is a community-supported system whose information is contributed by users. (The site will eventually make revenue from paid advertising on its web interface.) I was offered an early Freebase account several years ago, and I enjoy experimenting with and exploring the Freebase data. But I have not yet used Freebase in any work projects.

There are two ways that developers can use Freebase, which is based on the Metaweb database technology: as a source of structured data for real-world knowledge, and as a flexible data store for your own data. I will discuss only the first option. If you want a local copy of all Freebase data, you can get it from `http://download.freebase.com/datadumps/`. You license Freebase data using the Creative Commons Attribution (CC-BY) license. But you can use the examples in this section without creating a Freebase user account, because you'll only be reading existing Freebase data.

Freebase supports a query language called Metaweb Query Language (MQL), which uses JSON as a notation. You'll look at this notation, but you won't use it. Instead, you'll use a Ruby library that provides a more "Ruby-like" interface for querying Freebase.

I use the `freebase` Ruby gem developed by Christopher Eppstein, with contributions from Pat Allan. You need to install the `freebase` gem using `gem install freebase` (you must use Ruby 1.8.6). The `freebase` gem exposes Freebase domains as Ruby modules and Freebase types as Ruby classes. As an example, you'll see notation of the form `/business/advertising_slogan` for the Freebase domain "Business" and the Freebase type "Advertising slogan" within that domain. The `freebase` gem provides Ruby APIs to access domains and types. For the `business/advertising_slogan` example, the `freebase` gem uses published domains and types to allow access via `Business::AdvertisingSlogan`.

The easiest way to write Ruby scripts to fetch structured data from Freebase is to start with the web interface `http://www.freebase.com` and search for a domain and data type that you want to use. For example, if you want to find data types that you can access for businesses, enter "Business" in the Freebase search field. On the results page, click the first search result. The subsequent web page lists broad subtypes for business, such as "Industry" and "Advertising slogan." A sample MQL query might look like this (in JSON format):

```
query = { 'type': '/business/advertising_slogan',
          'name': 'It had to be good to get where it is.',
          'creator':[]}
```

This MQL query will match any business advertising slogan, and the value of the creator of this Freebase data entry will be returned (that is, it will match the empty array []).

The Ruby magic in the `freebase` gem dynamically builds Ruby modules and classes from type hierarchies found on `http://www.freebase.com`. Let's look at an example:

```
require 'rubygems'
require "freebase"
require 'pp'

a_slogan = Freebase::Types::Business::AdvertisingSlogan.find(:first)
pp "a_slogan:", a_slogan
puts "name=#{a_slogan.name}"
puts "creator =#{a_slogan.creator}"
```

The code in the gem generates the class name `AdvertisingSlogan` by "camel-casing" the business type "advertising_slogan." While it is possible to explore Freebase available types programmatically, I find that the best approach is finding useful types interactively by using the web application and then experimenting a little with the freebase gem to get the right Ruby module and class names to access the Freebase data. In the preceding code snippet, the output from pp a_slogan looks like this:

```
#<#<Freebase::Types::Business::AdvertisingSlogan brand:/business/brand_slogan>:➡
0x1934c60
 @result=
  {:type=>"/business/advertising_slogan",
   :timestamp=>
   [{"type"=>"/type/datetime", "value"=>Tue Feb 10 01:01:50 UTC 2009}],
   :permission=>
   [{"name"=>"Global Write Permission",
     "type"=>["/type/permission"],
     "id"=>"/boot/all_permission"}],
   :attribution=>
   [{"name"=>"skud",
     "type"=>["/type/user", "/type/namespace", "/freebase/user_profile"],
     "id"=>"/user/skud"}],
   :name=>"It had to be good to get where it is.",    ...
```

Printing a_slogan.name and a_slogan.creator produces the following output:

```
name=It had to be good to get where it is.
creator=/user/skud
```

As a last example, suppose you want to find all movie genres and all movie titles in each genre. From browsing http://www.freebase.com, I see that there is a domain called "Movie" and a type in this domain called "FilmGenre." Using the freebase gem module- and class-naming convention, you can find all genres using:

```
Freebase::Types::Film::FilmGenre.find(:all)
```

By pretty-printing the first genre result that's returned, you can see that there are accessor methods name and films_in_this_genre. The JSON hash table that is returned for films in a genre has a name key that you will also use. I get occasional runtime errors when calling the Freebase web services, so in this example I wrap the inner loop of web-service calls to Freebase in a begin/rescue block:

```
require 'rubygems' #required for Ruby 1.8.6
require "freebase"

all_genres = Freebase::Types::Film::FilmGenre.find(:all)
all_genres.each {|genre|
  puts "Movie genre: #{genre.name}"
  begin
    genre.films_in_this_genre.each {|film| puts "  #{film.name}"}
  rescue
    puts "Error: #{$!}"
```

```
    end
}
```

This example produces a lot of output; here is a snippet of the returned results:

```
Movie genre: Film noir
  Chinatown
  The Big Sleep
  Double Indemnity
  Blade Runner
  Gattaca
  Kiss Me Deadly
  The Third Man
```

I think that the Freebase system is an interesting idea—it's basically a community-driven site like Wikipedia but with structured data. I used a Ruby library to access Freebase in this section, but you can use other programming languages that have JSON libraries.

I used Open Calais in Chapter 3 to identify entity names on web pages. In the next section, you'll use Open Calais as a source of Linked Data.

Open Calais

I consider the Open Calais system (http://www.opencalais.com) to be the single most powerful and useful tool for working with Linked Data and the Semantic Web. You saw in Chapter 3 how you can use Open Calais to identify named entities such as people and places on web pages and other documents. When entities are identified, the Open Calais system attempts to define unique URIs that are also capable of being dereferenced. To dereference a URI is to perform an HTTP GET on it. Entity links generated by Open Calais reference resources at DBpedia.org, Wikipedia.org, Freebase.com, GeoNames.org, Shopping.com, IMDb.com, LinkedMDB.com, and Reuters.com.

The problem that Open Calais solves fairly well is dealing with different semantic meanings of entity names. The RDF response from web-service calls to analyze text often contains disambiguation information. I will show you a complete example for Ruby client code that uses Linked Data information (you can find the example files in the directory src/part3/open_calais_linked_data).

In Chapter 3, I used the XML comment block to easily get the entity names returned by Open Calais web-service calls. I will now use the RDF response. The goal in this example is to show you how to identify entities and their associated Open Calais disambiguated unique URIs. So, for example, if you were web-scraping many news stories, you could identify the entity "Barack Obama" as a unique URI across all of the news articles.

To keep this example simple, I will use input text containing people's names: "President George Bush and President Barack Obama played basketball." The Ruby script use_linked_data.rb redefines the utility class OpenCalaisTaggedText I developed in Chapter 3. The file sample.rdf contains the XML RDF payload that Open Calais returned after analyzing our test text. I converted this to N3 format to make it easier to read and understand:

```
cwm -rdf sample.rdf -n3 > sample.n3
```

If you look at the bottom of the file `sample.n3`, you see the three triples identifying Barack Obama. The subject for all three triples is the unique URI that Open Calais assigned to President Obama (I shortened these URIs to fit the line width):

```
<http://d.opencalais.com/pershash-1/cfcf1aa2-de05-3939-a7 ... 7b>
    a <http://s.opencalais.com/1/type/em/e/Person>;
    <http://s.opencalais.com/1/pred/name> "Barack Obama";
    <http://s.opencalais.com/1/pred/persontype> "political" .
```

There are many advantages of using the same URIs to refer to data from many different sources. If you want to find web pages that are similar to a page you are viewing, a smart browser might identify the most commonly mentioned entities (people, places, product names, businesses, and so on) on the web page and rank similar pages by how many of the same entities appear. This technique can produce better results than you'd get by simply matching stemmed words. For example, if you are reading a page about Paris, France, you do not want to see pages that contain people named Paris.

For the purposes of achieving our goal, I'll implement a new wrapper for the Open Calais web services that returns an array of entities with each entity represented by:

- The Open Calais URI for that entity
- The entity type (for example, "Person")
- The human-readable name for the entity (for example, "Barack Obama")

The first step is to develop XPath queries that extract this entity data for XML RDF. Here's the preceding N3 data, converted to the equivalent RDF XML (edited to fit page width):

```
<rdf:Description
  rdf:about='http://d.opencalais.com/pershash-1/cfcf1aa9-a7 ... 7b'>
    <rdf:type rdf:resource='http://s.opencalais.com/1/type/em/e/Person'/>
    <c:name>Barack Obama</c:name>
    <c:persontype> political </c:persontype>
</rdf:Description>
```

Now I'll rewrite the class `OpenCalaisTaggedText` from Chapter 3 and name it `OpenCalaisEntities` (refer to the file `use_linked_data.rb` for the complete implementation). It will look for specific URIs to determine the type of data in each `rdf:Description` element in the RDF data returned from the Open Calais web services. You need to do a `gem install simplehttp` if you do not have this gem installed:

```
require 'rubygems' # needed for Ruby 1.8.6
require 'simplehttp'
require "rexml/document"
include REXML
```

The `use_linked_data.rb` script expects you to have an Open Calais access key and have it defined in the environment variable `OPEN_CALAIS_KEY`. I shortened some of the header string constants here:

```
MY_KEY = ENV["OPEN_CALAIS_KEY"]
raise(StandardError,"Set Open Calais login key in ENV: 'OPEN_CALAIS_KEY'") ➥
if !MY_KEY

PARAMS = "&paramsXML=" + ... # see file use_linked_data.rb for complete string
```

The class constructor requires input text; it calls the Open Calais web services, parses the XML response, and saves the entity (person, place, and company) names. It also saves the unique Open Calais URI for each entity name:

```
class OpenCalaisEntities
  attr_reader :person_data, :place_data, :company_data
  def initialize text=""
    data = "licenseID=#{MY_KEY}&content=" + CGI.escape(text)
```

Make the web-service call to Open Calais:

```
    http = SimpleHttp.new "http://api.opencalais.com/enlighten/calais.asmx/➥
Enlighten"
    @response = CGI.unescapeHTML(http.post(data+PARAMS))
```

Define arrays to store results for people, places, and companies:

```
    @person_data = []
    @place_data = []
    @company_data = []
```

Here, I am using the XPath search support in REXML. REXML is inefficient for very large XML payloads, but I like using it to handle small XML documents. I am searching for every `rdf:Description` XML element:

```
    Document.new(@response).elements.each("//rdf:Description") {|description|
      about = name = type = nil
      uri = description.attributes['about']
      description.elements.each("rdf:type") {|e|
        # collect person data:
        type = 'Person' if e.to_s == "<rdf:type rdf:resource= ➥
                                  'http://s.opencalais.com/1/type/em/e/Person'/>"

        # collect place data:
        if e.to_s == "<rdf:type rdf:resource=      ➥
                                   'http://s.opencalais.com/1/type/er/Geo/City'/>" ||
           e.to_s == "<rdf:type rdf:resource=      ➥
                                   'http://s.opencalais.com/1/type/er/Geo/Country'/>" ||
           e.to_s == "<rdf:type rdf:resource=      ➥
                                   'http://s.opencalais.com/1/type/em/e/City'/>" ||
           e.to_s == "<rdf:type rdf:resource=      ➥
                                   'http://s.opencalais.com/1/type/em/e/Country'/>"
          type = 'Place'
        end
        # collect company data:
        if e.to_s == "<rdf:type rdf:resource=      ➥
```

```
                                    'http://s.opencalais.com/1/type/er/Company'/>" ||
            e.to_s == "<rdf:type rdf:resource=         ➥
                                    'http://s.opencalais.com/1/type/em/e/Company'/>"
          type = 'Company'
        end
      }
      description.elements.each("c:name") {|e| name = e.text if type}
      if uri && name
        @person_data << [uri, name] if type == 'Person'
        @place_data << [uri, name]  if type == 'Place'
        @company_data << [uri, name]  if type == 'Company'
      end
    }
  end
end
```

Note that in this example, I matched places and companies against multiple Open Calais resource types. The following code snippet uses the OpenCalaisEntities class:

```
tt = OpenCalaisEntities.new("President George Bush and President Barack Obama ➥
played basketball in London and in Paris France at IBM")

pp "people:", tt.person_data
pp "places:", tt.place_data
pp "companies:", tt.company_data
```

The output from this test is:

```
"people:"
[["http://d.opencalais.com/pershash-1/cfcf1aa2-de05-3939-a7d5-10c9c7b3e87b",
  "Barack Obama"],
 ["http://d.opencalais.com/pershash-1/63b9ca66-bfdb-3533-9a19-8b1110336b5c",
  "George Bush"]]
"places:"
[["http://d.opencalais.com/er/geo/city/ralg-geo1/➥
797c999a-d455-520d-e5cf-04ca7fb255c1",
  "Paris,France"],
 ["http://d.opencalais.com/er/geo/city/ralg-geo1/➥
f08025f6-8e95-c3ff-2909-0a5219ed3bfa",
  "London,Greater London,United Kingdom"],
 ["http://d.opencalais.com/er/geo/country/ralg-geo1/➥
e165d4f2-174b-66a7-d1a9-5cb204d296eb",
  "France"],
 ["http://d.opencalais.com/genericHasher-1/56fc901f-59a3-3278-addc-b0fc69b283e7",
  "Paris"],
 ["http://d.opencalais.com/genericHasher-1/e1fd0a20-f464-39be-a88f-25038cc7f50c",
  "France"],
 ["http://d.opencalais.com/genericHasher-1/6fda72fd-105c-39ba-bb79-da95785a249f",
```

```
  "London"]]
"companies:"
[["http://d.opencalais.com/er/company/ralg-tr1r/➡
9e3f6c34-aa6b-3a3b-b221-a07aa7933633",
  "International Business Machines Corporation"],
 ["http://d.opencalais.com/comphash-1/7c375e93-de13-3f56-a42d-add43142d9d1",
  "IBM"]]
```

If you have signed up for a free Open Calais license, you can make 40,000 web-service calls per day at a maximum rate of four transactions per second. If this is insufficient for your use, you can purchase a fee-based plan.

Wrapup

You began this chapter with a tutorial about using the D2R system, which turns relational databases into Linked Data sources. Specifically, D2R wraps databases to look like virtual RDF data stores with a SPARQL interface. D2R does not actually convert your data to RDF triples for storage, but converts relational data to triples on the fly when that data is requested.

You also looked at the Linked Data sources DBpedia, Freebase, and Open Calais. You can use these Linked Data sources directly in your applications, and they also serve as implementation examples for your own publication systems.

In Chapter 12, I will cover strategies for information storage. Getting systems running locally in test mode is a first step, but development is not complete until you work out how to deploy your applications in a cost-effective way that can also scale to large numbers of users.

Implementing Strategies for Large-Scale Data Storage

I've been discussing different types of data storage. But what happens when you have too much data to be hosted on a single server using a relational database or RDF data store? Or what happens if you need to speed up data read/write operations because the latency is too large for users of your web applications?

I begin this chapter mentioning ways to scale PostgreSQL and MySQL databases. I won't spend too much time on these scaling options because relevant implementations are well-documented on the Web. I'll spend most of this chapter on alternatives to relational databases that you might not have seen or used, including memcached, CouchDB, and Amazon S3. I end this chapter with a look at Amazon EC2 server on-demand services.

Note Appendix A discusses an Amazon EC2 AMI that contains most of the examples in this book, configured and ready to run.

Using Multiple-Server Databases

The easiest way to deploy large-scale relational databases is to use a single server with sufficient CPU, memory, and disk resources to handle your application with the number of simultaneous users that you expect to have. For cases when a single database server is insufficient, you might want to consider using multiple servers. I will review two multiple-server strategies: master/slave data replication, and database sharding.

Database Master/Slave Setup for PostgreSQL

If most access to a database consists of read operations, then an effective way to scale is to implement a *master/slave* setup. You maintain one *master* database for all write and update operations, and several *slave* database servers that stay in sync with the master and process database read operations. Slave servers stay in sync by reading the master's journal or log files.

There are several open source packages that support master/slave PostgreSQL systems. Probably the most popular is Slony-I (http://www.slony.info), so I'll cover that option here.

To install Slony-I, you should download the latest source code bundle and also download a source code bundle of PostgreSQL that is compatible with Slony-I (as determined in the Slony-I INSTALL file). Build and install PostgreSQL according to the directions on http://www. postgresql.org. You then need to determine where the pg_config program is; on my system the location is /usr/local/pgsql/bin, so I configured Slony-I using:

```
cd slony1-2.0.1
./configure --with-pgconfigdir=/usr/local/pgsql/bin
make
sudo make install
```

Once you have installed Slony-I, I recommend that you set up a new test database and configure just one master and one slave PostgreSQL service using the directions at http:// www.slony.info/documentation/. Installing a PostgreSQL master/slave cluster using Slony-I is not easy. The first time I set up a test cluster on the same server (the easiest way to experiment with Slony-I and PostgreSQL), it took me several hours. That said, the documentation is very thorough and reasonably easy to follow.

Database Master/Slave Setup for MySQL

Replication of data from a master server to slave servers for MySQL is similar to the corresponding process in PostgreSQL. If you have MySQL 5.x installed on the server used as the master and on the server(s) used as slaves, then you need to perform just a few steps to configure a replicated database. You will want to refer to the documentation at http://dev.mysql. com/doc/refman/5.0/en/replication.html.

Each MySQL slave process needs to connect to the master process using a standard MySQL account and password, so you need to grant slave-process access on your server instance:

```
GRANT REPLICATION SLAVE ON *.* TO 'repl'@'%.masterserver_domain.com'   ➥
          IDENTIFIED BY 'password for slaves';
```

Replication occurs when the slaves read a binary log file produced by the master. You need to configure the master setup (the /etc/my.cnf file on Linux, Unix, and Mac OS X, or the my.ini file on Windows) to include:

```
[mysqld]
log-bin=mysql-bin
server-id=1
```

When you set up a slave, you want the server-id to be both unique and greater than the server-id for the master. As an example, you will want to add the following to the slave /etc/ my.cnf file:

```
[mysqld]
server-id=2
```

In general, when you first start performing data replication, you need to follow the steps in the MySQL documentation for stopping the master process and dumping the data to the slave(s). To make experimenting with MySQL replication easier, I suggest that you start with a new empty database that contains an empty test table.

You do not need to dump this new database using `msqldump` and import this dump file on the slave(s) because we are assuming that the new database to be replicated is empty. On each slave, you need to start the service and execute the `CHANGE MASTER TO` command to set the master server's IP address, slave account name, password, and so on (see `http://dev.mysql.com/doc/refman/5.0/en/change-master-to.html` for reference).

Even when a single database server can handle the load for a web application, you can still use replication to perform a "hot backup" of your database. Having a single master database server might not scale to your workload if you have many write operations; remember that all write operations occur on the master. But the slaves can handle read requests from your web application.

In addition to the performance gains that you get from using database replication, you enjoy other advantages. When developing new versions of software, you might find it beneficial to use "real data" for testing. You can accomplish this using replication: specifically, you can synchronize one slave process to the master server and then temporarily disconnect it and use it for write, update, and read operations. You can also use replication in data warehousing and offline analytics that require dedicated use of a database for non–real-time computations.

Database Sharding

Some database applications cannot be scaled out using master/slave replication because the rate of write and update operations is too high for a single server. In this case, you can use *sharding* to perform the scaling. Database sharding assumes that you can break up databases based on some criteria that determines the table rows that will go to specific servers. Sharding can also involve partitioning entire tables to different servers. Implementing sharding is a lot of work, and you should avoid it if possible because the use of sharding strongly affects system architecture and implementation.

■**Caution** You usually want to normalize your database schemas by eliminating repeating groups of attributes in a table, by using key attributes to associate multiple rows in a table with a single row in another table, and by generally avoiding redundant data. Normalized database schemas make high-volume transactions more efficient. However, the combination of using sharding with highly normalized database schemas can be very inefficient. *Not* using normalized schemas and using sharding can make sense in cases where data is archival in nature—that is, if it's seldom modified once it's written to the database. Sharding can be very effective when you have a high volume of write operations.

MySQL does support user-defined partitioning. When you build MySQL from source, add `--with-partition` when you run `./configure` (see `http://dev.mysql.com/doc/refman/5.1/en/partitioning.html` for reference). For PostgreSQL, you might want to look at the `pgpool` project (`http://pgpool.projects.postgresql.org/`) as a starting point for implementing sharding.

Caching

Caching is the temporary storage of precomputed results. For effective use, the time spent looking up data for given a key value in a cache should be much less than recalculating the data.

One way to avoid (or postpone) the need for implementing master/slave data replication or database sharding is to cache the results of SQL queries. Before submitting a query to a database server, look the query up in a cache. If the query is in the cache and the results were calculated recently, then return the cached query results (the definition of "recent" is a tunable parameter). If the results in the cache are too old, clear the cache entry and submit the query to the database. Whenever results are fetched from a database, add them to the cache.

Using memcached

Now I'll delve into the first alternative to using a relational database: memcached. memcached is an efficient and easy-to-use distributed caching system that I use to cache web-service calls and database queries, and to handle user sessions in web applications. You can download the source code for memcached at `http://www.danga.com/memcached/`. Before you install memcached, you need to install libevent (`http://www.monkey.org/~provos/libevent/`). memcached is used on Slashdot, Twitter, Facebook, and other high-volume web sites.

The example programs for this section are in the directory `src/part3/memcached` (you can find the code samples for this chapter in the Source Code/Download area of the Apress web site at `http://www.apress.com`). In the following discussion, I will assume that you have memcached installed. If you don't already have it installed, follow the directions at `http://www.danga.com/memcached/`. I will use the memcached client gem written by Michael Granger, so you need to `gem install Ruby-MemCache`.

By default, the memcached service runs on port 11211. For the next example, I am going to use just one instance on the default port. If you need more than one memcached process, you can use the `-p` option to override the default port. For reference, here is an example of starting two memcached processes and running them as daemons:

```
memcached -d              # default port is 11211
memcached -p 11212 -d
```

The example program `memcached_client_example.rb` uses a single memcached process on localhost. In the following output, I am running this example script in `irb` so that you can see the output:

```
>> require 'memcache'
=> true
>> require 'pp'
=> false
>> cache = MemCache::new('localhost:11211',
?>                                      :debug => false,
?>                                      :namespace => 'ruby_test')
=> MemCache: 1 servers buckets: ns: "ruby_test", debug: false, cmp: true, ro: false
```

The next statement adds an array containing one string with a key value of "Ruby":

```
>> cache['Ruby'] = ["http://www.ruby-lang.org/en/"]
=> ["http://www.ruby-lang.org/en/"]
>> pp cache['Ruby']
["http://www.ruby-lang.org/en/"]
=> nil
```

If I want to add another string to the array associated with the key "Ruby," I need to read from the cache, update the data, and write back to the cache:

```
>> cache['Ruby'] = cache['Ruby'] << "http://www.ruby-lang.org/en/libraries/"
=> ["http://www.ruby-lang.org/en/", "http://www.ruby-lang.org/en/libraries/"]
>> pp cache['Ruby']
["http://www.ruby-lang.org/en/", "http://www.ruby-lang.org/en/libraries/"]
=> nil
```

Any keys that are symbol values are converted to strings:

```
>> cache[:a_symbol_key] = [1, 2, 3.14159]
=> [1, 2, 3.14159]
>> pp cache['a_symbol_key']
[1, 2, 3.14159]
=> nil
```

This conversion to strings makes sense because symbols are unique in a Ruby runtime environment, but unique symbols cannot be shared between multiple memcached clients and a memcached server.

Values added to memcached can be any Ruby object that can be serialized. The memcached gem handles this serialization automatically:

```
>> cache["an_array"] = [0, 1, 2, {'cat' => 'dog'}, 4]
=> [0, 1, 2, {"cat"=>"dog"}, 4]
>> value = cache["an_array"]
=> [0, 1, 2, {"cat"=>"dog"}, 4]
>> value[3]
=> {"cat"=>"dog"}
>> p cache['no_match'].class
NilClass
=> nil
```

In this example, I connected to a single memcached server. But you can also specify multiple servers and optional arguments to compress data before storing it in memcached. Read the complete API documentation at http://www.deveiate.org/code/Ruby-MemCache/.

Using memcached with ActiveRecord

There are a few projects that provide memcached wrappers for ActiveRecord. However, because you can implement a wrapper in about a dozen lines of code, in this example I'll implement a wrapper from scratch. Because I am using an automatic cache-entry expiration in this example, you will need to install gem install system_timer.

I start by requiring both the ActiveRecord and memcached gems:

```
require 'rubygems' # require for Ruby 1.8.6
require 'activerecord'
require 'memcache'
```

And here I'm making a connection to the same MySQL database that I used in Chapter 8:

```
ActiveRecord::Base.establish_connection(
  :adapter  => :mysql,
  :database => "test"
)
```

Next, I derive a class from ActiveRecord::Base that wraps the database table news_articles. I create a class attribute @@cache that makes a connection to a local memcached service. The new class method find_cache first checks whether the cache contains a news article with the specified ID. If no such article exists in the cache, find_cache reads the object from the MySQL database. Whenever an object is read from the database, it is saved in the cache. The method update_cache calls the class method update to save the object to the database and then updates the copy of the object in the cache:

```
class NewsArticle < ActiveRecord::Base
  @@cache = MemCache::new('localhost:11211', :namespace => 'news_articles')
  def self.find_cache id
    article = @@cache[id]
    if !article
      puts "id #{id} not in cache"
      article = NewsArticle.find(id)
```

Instead of using the Ruby hash [] operators, I am using the set method. This method takes an optional third argument, which is the number of seconds that a key is valid before it expires automatically. Here I set the expiration time to 180 seconds:

```
      @@cache.set(id, article, 180)
    end
    article
  end
  def update_cache
    update # call class method
    @@cache.set(self.id, self, 180)
  end
end
```

The following examples show how the news articles are initially loaded from the database but are then available from the memcached service. Here, I change the URL of a news article and update both the database and the cache:

```
article_1 = NewsArticle.find_cache(1)  # not in cache
article_2 = NewsArticle.find_cache(2)  # not in cache
puts "#{article_1.url} #{article_2.url}"


article_1 = NewsArticle.find_cache(1)  # now in cache
article_2 = NewsArticle.find_cache(2)  # now in cache
puts "#{article_1.url} #{article_2.url}"


article_1.url = "http://test.com/bigwave" + rand(999).to_s
article_1.update_cache


article_1 = NewsArticle.find_cache(1)
puts "modified url for first article: #{article_1.url}"
```

The following output shows this example run twice to illustrate the effect of caching:

```
$ ruby activerecord_wrapper_example.rb
id 1 not in cache
id 2 not in cache
http://test.com/bigwave645 http://test.com/bigfish
http://test.com/bigwave645 http://test.com/bigfish
modified url for first article: http://test.com/bigwave64
$ ruby activerecord_wrapper_example.rb
http://test.com/bigwave64 http://test.com/bigfish
http://test.com/bigwave64 http://test.com/bigfish
modified url for first article: http://test.com/bigwave496
```

Notice in this output that once a news article identified by a key value equal to its database row ID is in the cache, the database is not accessed. memcached will automatically delete key/value pairs when it runs out of storage, deleting keys that have not been recently accessed. In the last example, automatically expiring keys after 180 seconds incurred overhead because the local Ruby client must keep keys in (local) memory and set a timer to expire them.

Using memcached with Web-Service Calls

If web-service calls take significant server resources to process, then it can make sense to cache web-service results. The example in this section uses memcached with a SPARQL client example from Chapter 11 to make queries against DBpedia.

The strategy used in this example is simple: cache query results using a key that consists of the last 255 characters of the SPARQL query. Note that memcached has a limit on key lengths of 256 bytes. In this example, we can expect the last 255 characters to include all of the actual query, but not necessarily all of the prefix statements. A more complicated implementation that you might have to use for very long keys is to calculate a shorter cryptographic hash for queries and limit the hash size to 255 characters.

I am including the benchmark gem so that you can compare the query runtime using SPARQL and memcached:

```
require "rubygems"
require "sparql_client"
require 'memcache'
require 'benchmark'
```

Notice that I am specifying an application-specific namespace for memcached:

```
CACHE = MemCache::new('localhost:11211', :namespace => 'dbpedia')
```

This technique uses memcached similarly to the way I used it with ActiveRecord in the preceding section, except that I do not expire cache entries after a specific time interval because I assume that the DBpedia data is fairly static:

```
def sparql_query_cache endpoint, query
  key = endpoint + query
  klength = key.length
  puts "Key length = #{key.length}"
  results = CACHE[key[klength - 255..-1]]
  if !results
    puts "SPARQL query to DBPedia not in cache"
    sparql = SPARQL::SPARQLWrapper.new(endpoint)
    sparql.setQuery(query)
    results = sparql.query
    CACHE[key[klength - 255..-1]] = results
  end
  results
end
```

I now specify a query string and a SPARQL endpoint and make two identical queries:

```
qs="
SELECT distinct  ?pred WHERE {
     ?a ?pred ?o .
}
LIMIT 100
"
endpoint="http://dbpedia.org/sparql"

# Make the same query twice to show effect of caching:

2.times {|i|
  puts "Starting SPARQL DBPedia query..."
  puts Benchmark.measure { puts sparql_query_cache(endpoint, qs) }
}
```

The output shows that the first call takes 0.66 seconds while the second call takes 0.03 seconds:

```
#<SPARQL::QueryResult:0x1282c28>
  0.090000   0.010000   0.100000 (   0.663204)
Starting SPARQL DBPedia query...
Key length = 88
#<SPARQL::QueryResult:0x11af38c>
  0.030000   0.010000   0.040000 (   0.037053)
```

memcached is a great tool for scaling Web 3.0 applications. I covered the use of memcached for caching results from database queries and web-service calls. Although I don't cover it here, memcached can also make Rails applications more efficient if you use it to store user HTTP session data and controller actions.

The next section introduces the CouchDB distributed-caching system for document storage.

Using CouchDB

CouchDB is a distributed data store for documents. Documents are identified by unique keys that are assigned as documents are added. For our purposes, you can consider a document to be any data represented in JSON. And because you'll be using a Ruby CouchDB client, the JSON values can be any Ruby objects that can be serialized. Internally, CouchDB saves documents in JSON format.

CouchDB provides several advantages over relational databases that can also be disadvantages, depending on your application. Data schemas in CouchDB are ad hoc: you can add attributes to documents. Documents are also self-contained—not like normalized databases that split data over multiple tables to reduce duplication and prevent inconsistent data updates. CouchDB is written in Erlang (with a JavaScript interpreter written in C). Erlang was designed for writing distributed, fault-tolerant applications, and CouchDB is designed to facilitate data replication over multiple servers. CouchDB does not provide a query language like SQL, but it does allow custom "views" to transform data and return results from queries.

CouchDB views are implemented in JavaScript as paired *map/reduce functions*, which are saved with their associated databases. Map/reduce functions operate on key/value pairs. Map functions transform key/value pairs using a one-to-one mapping, while reduce functions perform many-to-one transformations. I will be using map/reduce functions in Chapter 14 when I discuss large-scale, distributed data processing.

I am not covering in any detail the use of internal views implemented as map/reduce functions written in JavaScript. A view transforms data by sorting it, extracting specific data for client applications, and so on. For information about writing JavaScript views that run inside CouchDB processes, refer to the free O'Reilly Media book *CouchDB: The Definitive Guide* at http://books.couchdb.org/relax/ as well as the CouchDB documentation at http://couchdb.apache.org/. If you want to experiment with views' map/reduce functions, there are code samples in the examples directory within the couchrest gem installation directory.

One of the gem examples in couchrest/examples/word_count loads four large text files into CouchDB and defines the following map/reduce functions in JavaScript to count the total number of words in all documents:

```
word_count = {
  :map => 'function(doc){
    var words = doc.text.split(/\W/);
    words.forEach(function(word){
      if (word.length > 0) emit([word,doc.title],1);
    });
  }',
  :reduce => 'function(key,combine){
    return sum(combine);
  }'
}
```

This hash table has two keys, :map and :reduce, the values of which are JavaScript functions that are executed inside the CouchDB process when the word_count view is requested by a client. So, map functions get called for every document, but they can ignore documents based on various criteria such as document type, existence of an attribute, a custom search of data internal to a document, and so on. If a map function decides *not* to ignore a document, then it can call a special emit function with any calculated data as arguments to emit. The reduce function collects all "emitted" data and returns it as the reduce function's value. Map and reduce functions for views are stored inside of and associated with specific CouchDB data stores.

CouchDB is much more than a document store; it is a platform for web-based applications running JavaScript views and map functions on the server to support HTML/JavaScript web-browser clients. JavaScript views' map and reduce functions cache their results after they are run and get called again only when the underlying document data changes. Cached view and map results are stored in a B-Tree storage. CouchDB is an amazing project, and I hope that this short introduction will encourage you to experiment with it on your own.

■**Note** JavaScript is commonly used on the browser side of web applications, but lately there has also been increased use of JavaScript on the server side. Google, Apple, and the Mozilla group are all working on much faster JavaScript engines. This research and development can only help the development of CouchDB, which internally uses JavaScript.

In order to follow along with the examples in this chapter, you will need to install Erlang (http://erlang.org) and CouchDB (http://couchdb.apache.org/). CouchDB has a dependency on the SpiderMonkey JavaScript engine, so install that first (https://developer. mozilla.org/en/SpiderMonkey). There are also prebuilt versions of CouchDB for some operating systems.

■**Tip** Are you having difficulty setting up CouchDB? You can use my Amazon EC2 AMI instance that has the book examples already up and running (see Appendix A). If you want to try the JavaScript view examples that ship with the gem, the location of the couchrest gem on my AMI is /var/lib/gems/1.8/gems/ couchrest-0.22.

I use the couchrest Ruby gem, written by Chris Anderson, that you can install in Ruby 1.8.6 using gem install couchrest. I am not covering the topic of JavaScript CouchDB views and mapping functions here, but the examples directory in the couchrest gem has several examples that you can experiment with. The first example (couchdb_test.rb) is:

```
require 'rubygems' # required for Ruby 1.8.6
require 'couchrest'
require 'pp'

##  from web site documentation (and gem README file):

# with !, it creates the database if it doesn't already exist
db = CouchRest.database!("http://127.0.0.1:5984/test")
response = db.save_doc({'key 1' => 'value 1', 'key 2' => ➥
[1, 2, 3.14159, 'a string']})
pp "response:", response

doc = db.get(response['id'])
pp doc

# returns ids and revs of the current docs
pp db.documents
```

If you have just started the CouchDB service or used the web interface (see Figure 12-1) to delete the test database for the purposes of rerunning this example, then the output of this code snippet is:

```
$ ruby couchdb_test.rb
"response:"
{"rev"=>"1476925524", "id"=>"04113db8f2bdc1499f7b820e547c996d", "ok"=>true}
{"_id"=>"04113db8f2bdc1499f7b820e547c996d",
 "_rev"=>"1476925524",
 "key 1"=>"value 1",
 "key 2"=>[1, 2, 3.14159, "a string"]}
{"rows"=>
  [{"id"=>"04113db8f2bdc1499f7b820e547c996d",
    "value"=>{"rev"=>"1476925524"},
    "key"=>"04113db8f2bdc1499f7b820e547c996d"}],
 "offset"=>0,
 "total_rows"=>1}
```

CouchDB provides an administration web interface that you can access using the URL http://127.0.0.1:5984/_utils/ (see Figure 12-1).

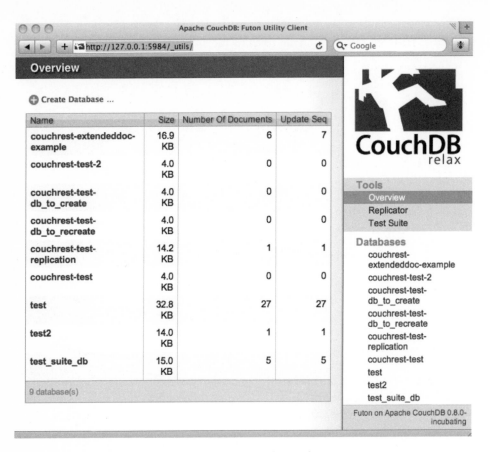

Figure 12-1. *Using the CouchDB administration web interface*

Saving Wikipedia Articles in CouchDB

We will reuse TextResource (and its derived classes) from Chapters 1, 2, and 3 to extract semantic information from Wikipedia articles and save this information in CouchDB. The directory src/part3/couchdb_client/test_data contains the text for two Wikipedia articles that we will use in this section.

The file wikipedia_to_couchdb.rb takes the sample Wikipedia articles that have been saved as local files and uses the TextResource classes to save the extracted semantic data to CouchDB. The CouchDB service returned an error when I tried to store arrays that hold many strings containing person and place names, so I encoded these arrays using the data-serialization language YAML:

```
require 'rubygems' # required for Ruby 1.8.6
require 'couchrest'
require 'yaml'
require 'text-resource'

require 'pp'
```

```
tr1 = PlainTextResource.new('test_data/wikipedia_Barack_Obama.txt')
tr2 = PlainTextResource.new('test_data/wikipedia_Hillary Rodham Clinton.txt')
```

I am going to store the semantic data from these two articles in a CouchDB data store named wikipedia_semantics:

```
db = CouchRest.database!("http://127.0.0.1:5984/wikipedia_semantics")
```

```
[tr1, tr2].each {|tr|
  response = db.save_doc({
    'source_uri' => tr.source_uri,
    'summary' => tr.summary,
    'sentiment_rating' => tr.sentiment_rating.to_s,
    'human_names' => YAML.dump(tr.human_names),
    'place_names' => YAML.dump(tr.place_names),
    'plain_text' => tr.plain_text
    })
}
```

The output from running this code snippet consists of a list of CouchDB IDs that are added to the wikipedia_semantics data store:

```
doc id: 7acd61d493562aeca64c248a6b33b320
doc id: 31dd78ed932ad76dda9e2fab1672e5cf
```

We'll use these CouchDB document IDs in the next example program, which will read the article information from CouchDB.

Reading Wikipedia Article Data from CouchDB

Now you'll work through an example program that reads the CouchDB data store wikipedia_semantics you created in the preceding section. I use the CouchRest method documents to get the IDs of the documents in this data store (the following code is in the file wikipedia_from_couchdb.rb):

```
require 'rubygems' # required for Ruby 1.8.6
require 'couchrest'
require 'yaml'
require 'text-resource'

require 'pp'

db = CouchRest.database!("http://127.0.0.1:5984/wikipedia_semantics")
ids = db.documents['rows'].collect {|row| row['id']}

ids.each {|id|
  doc = db.get(id)
  puts doc['source_uri']
  puts doc['summary']
```

```
  puts doc['sentiment_rating']
  pp YAML.load(doc['human_names'])
  pp YAML.load(doc['place_names'])
  puts doc['plain_text']
}
```

The output from this code snippet is a few hundred lines long, so I am not showing it here. I hope that you are working along with this example and that you've tried running these two programs:

```
$ ruby wikipedia_to_couchdb.rb
$ ruby wikipedia_from_couchdb.rb
```

The simplest use of CouchDB is for storing structured data in a high-performance local storage system. However, CouchDB can serve as a complete development and deployment environment if you use the Futon web application seen in Figure 12-1. Even though I have used CouchDB only as a simple distributed-storage tool in this chapter, I would like to leave you with the idea that CouchDB can be a complete platform for developing web applications and services.

If you decide to use CouchDB as a development platform, install CouchApp, the Python tools for application development (using the command sudo easy_install couchapp). As I write this chapter, CouchApp is under active development and has just been converted from Ruby to Python. Search for "CouchApp" on the Web for tutorials.

You can run many instances of CouchDB on a local network to scale applications that need to perform frequent data-store read/write operations. In the next section, we will look at a similar scalable data store for structured data: Amazon S3.

Using Amazon S3

Amazon Simple Storage Service (S3) is a low-cost way to store large amounts of data on Amazon's servers. You create, modify, read, and remove data in S3 using web-service calls. I will use S3 in Chapter 14 both to provide data for large-scale, distributed data-crunching tasks and to provide storage for calculated results. Also in Chapter 14, I will show you how to use command-line tools to manipulate data in S3. Here, I cover using S3 in Ruby programs.

In order to use any of the Amazon Web Services (AWS), you need to create an account at http://aws.amazon.com/. You will be assigned both an Access Key ID and a Secret Access Key. I define environment variables that set these values, so I don't have to put my own login information in my code:

```
export AMAZON_ACCESS_KEY_ID="YOUR ACCESS ID"
export AMAZON_SECRET_ACCESS_KEY="YOUR SECRET KEY"
```

For each AWS account, you can define separate S3 namespaces. Namespaces are called *buckets*, and the name of each one must be unique across all of S3.

The Ruby script I'll illustrate in this section, s3_test.rb, performs read and write operations of strings, binary files, and Ruby objects to and from S3 (you can find the script in this book's source code, in the src/part3/amazon-s3 directory). To try it out, you need to install Marcel Molina, Jr.'s gem for accessing S3: gem install aws-s3.

The following output is generated when you load s3_test.rb into irb:

```
$ irb
>> require 'aws/s3'
=> true
>> require 'yaml'
=> false
>> require 'pp'
=> false
```

I start by making a connection to the S3 service (edited to fit line width):

```
>> AWS::S3::Base.establish_connection!(
?>            :access_key_id => ENV['AMAZON_ACCESS_KEY_ID'],
?>            :secret_access_key => ENV['AMAZON_SECRET_ACCESS_KEY'])
=> #<AWS::S3::Connection:0x1cfc190    ...>
```

If I print out the buckets that I have in my account, you will see the S3 storage for my Amazon AMI that contains the book examples as a ready-to-run Linux image (see Appendix A):

```
>> pp AWS::S3::Service.buckets
[#<AWS::S3::Bucket:0x1cec920
  @attributes=
   {"name"=>"markbookimage", "creation_date"=>Sat Mar 28 19:53:47 UTC 2009},
  @object_cache=[]>]
=> nil
```

Next, I create a test bucket named web3_chapter12 to show you how to put strings, binary data, and Ruby objects into S3 storage:

```
>> pp AWS::S3::Bucket.create('web3_chapter12')
true
=> nil
```

You use the find method to get a local interface for a bucket:

```
>> examples_bucket = AWS::S3::Bucket.find('web3_chapter12')
=> #<AWS::S3::Bucket:0x1cdec94 @object_cache=[], @attributes={"prefix"=>nil, ⮞
"name"=>"web3_chapter12", "marker"=>nil, "max_keys"=>1000, "is_truncated"=>false}
```

Next, I store the binary data from a PDF file in S3:

```
>> AWS::S3::S3Object.store('FishFarm.pdf',
?>         open('/Users/markw/Sites/opencontent/FishFarm.pdf'), 'web3_chapter12')
=> #<AWS::S3::S3Object::Response:0x15110200 200 OK>
>> pp examples_bucket.objects
[#<AWS::S3::S3Object:0x15096270 '/web3_chapter12/FishFarm.pdf'>]
=> nil
```

And now I can read the binary data for this test PDF file back from S3:

```
>> pdf = AWS::S3::S3Object.find('FishFarm.pdf', 'web3_chapter12')
=> #<AWS::S3::S3Object:0x15074650 '/web3_chapter12/FishFarm.pdf'>
```

If I wanted access to the data, I could use the attribute value, referencing pdf.value. For binary data that I want to read from S3 and copy to a local file, I can use the method stream:

```
>> open('/tmp/FishFarm.pdf', 'w') do |file|
?>    AWS::S3::S3Object.stream('FishFarm.pdf', 'web3_chapter12') {
?>                          |data| file.write(data) }
>> end
=> "#<Net::ReadAdapter:0x1cb3530>"
```

The preceding code snippet writes the binary contents of the PDF file to my local file system as /tmp/FishFarm.pdf. In order to store the value of Ruby data, I encode the data using the YAML class:

```
>> AWS::S3::S3Object.store('test ruby data 1',
?>                                    YAML.dump([1,2,3.14159,"test"]), ➥
'web3_chapter12')
=> #<AWS::S3::S3Object::Response:0x15020210 200 OK>
```

The method find returns the YAML-encoded string value:

```
>> data = AWS::S3::S3Object.find('test ruby data 1', 'web3_chapter12')
=> #<AWS::S3::S3Object:0x15001270 '/web3_chapter12/test ruby data 1'>
```

Here, I decode the YAML string back to a Ruby object:

```
>> pp YAML.load(data.value)
[1, 2, 3.14159, "test"]
=> nil
```

The next two statements delete the two test objects from the web3_chapter12 bucket:

```
>> AWS::S3::S3Object.delete('test ruby data 1', 'web3_chapter12')
=> true
>> AWS::S3::S3Object.delete('FishFarm.pdf', 'web3_chapter12')
=> true
```

Converting the examples in the preceding section to use S3 would be trivial. That is, you could write structured information from Wikipedia articles to S3 instead of CouchDB, and then read the data; I leave that as an exercise for you.

Because CouchDB and S3 provide similar services, modifying applications to use either CouchDB or S3 shouldn't be too difficult. But S3 does not support server-side scripting (JavaScript views and map/reduce functions) as CouchDB does, so you might want to avoid using CouchDB's server-side scripting capabilities if you plan to write your applications to use either back-end storage system.

Amazon S3 is a low-cost storage solution, and more than a few web applications use S3 rather than implement their own storage on local servers. S3 is also very appropriate for implementing off-site backup for applications running on your own servers. In the next section, I will introduce you to EC2, another Amazon service that works well with S3.

Using Amazon EC2

Amazon Elastic Compute Cloud (Amazon EC2) is a web service that provides resizable compute capacity that you can modify in almost real time to adjust to increased (or decreased) traffic on your web applications. EC2 *server instances* are virtual servers that are available with different CPU and memory options. You create custom operating-system images to run on server instances that you rent by the hour. I always use Linux on my EC2 server instances, but Windows is also available. If you rent a server instance full time (24/7), the price is roughly the same as renting a raw server or a virtual private server (VPS) with the same capabilities. Operating-system images that are loaded with desired application software are called Amazon Machine Images (AMIs).

■Note Amazon EC2 services compete with VPS-hosting companies such as RimuHosting and Slicehost. In my work, I use semi-managed VPSs for deploying customer web portals that do not need to support high-volume traffic. I have had one customer portal (written in Java) run for four years unattended on a VPS (except for backups): "set it and forget it" deployment. EC2 is more appropriate when you must design your systems to run on multiple servers and you need to be able to quickly scale up or down by adding or removing servers. EC2 is implemented using VPS technology.

In Appendix A, you will learn how to use an AMI that has almost all of the example programs and services from this book. Renting a server instance and loading a prebuilt AMI costs about 10 cents an hour, so you can experiment fairly inexpensively.

If you followed along with the S3 examples in the preceding section, then you already have an account; if not, you might want to create one now (http://aws.amazon.com/). This chapter deals specifically with strategies for scalable storage for Web 3.0 applications, but EC2 offers more general capabilities in addition to providing a storage solution. When you set up an AMI for a web application, you can start up any number of instances (remember, however, that you are paying by the hour for each instance).

As an example, suppose you had a web application that used Ruby on Rails, a PostgreSQL database, and a Sesame RDF data-store web service running on Tomcat. If you had relatively few users, you could run all of these processes on a single server. As the number of your users increases, you can almost instantly upgrade to a server instance with more CPU and memory resources, and start running your different processes on multiple server instances as necessary. If you offload your back-end storage to S3, you might find that a single server instance can handle a large number of simultaneous users (depending on your application).

Another great use of EC2 is as a learning platform. If you want to practice writing scalable applications and configuring multiple cooperating servers, for example, you can rent three small server instances for an eight-hour workday for less than three dollars. I wouldn't be surprised if schools and universities were to start using EC2 for student learning projects.

Amazon supplies a web-based administration console called the AWS Management Console (https://console.aws.amazon.com/), which you can use to manage AMIs, start and stop server instances, and so on. Amazon provides great online documentation that you should read before getting started.

The first time that you try using EC2, you will use the AWS Management Console to choose an existing AMI that is likely to be configured with the infrastructure software you need. Figure 12-2 shows an example where I have chosen to see only public AMIs that run Ubuntu Linux and that have Ruby on Rails already installed.

Figure 12-2. *Using the AWS Management Console to find a public Ubuntu Linux AMI that is set up for Ruby on Rails*

■**Caution** When you are done using server instances, make sure that you stop them. You are billed in one-hour increments for the use of each server instance.

The AWS Management Console allows you to allocate and control resources. However, you can also install command-line tools on your development system for managing AWS assets. I find that I mostly use the command-line tools, but the web-based management console has a much smaller learning curve when you are just getting started.

Elastic IP addresses are static IP addresses associated with your AWS account, and you can allocate them to any server instances that you currently have running. This allows you to implement hot failovers to backup server instances. Also, if your web application uses more than one server instance, you can assign an Elastic IP address to each server for network connections.

Wrapup

When developing public Web 3.0 applications, you should plan ahead for scaling your application as the number of your users increases. It's not that difficult to scale for more users with infrastructure like Ruby on Rails or a Java web container (Apache Tomcat, JBoss, IBM WebSphere, Sun GlassFish, and so on), but you face the problem of maintaining HTTP session data for many simultaneous users. However, scaling data storage can be more difficult. Using a service like S3 provides scalable storage but offers less flexibility than using relational databases, RDF data stores, or a customizable document-storage platform such as CouchDB.

One interesting scaling option is the Google App Engine (http://code.google.com/appengine/). Currently, App Engine supports the Python and Java programming languages, and I run two of my web applications using the Java version of App Engine. I have not covered App Engine in this book because I'm focusing primarily on Ruby applications; I did not want to provide long examples written in Python or Java. It is possible to run Rails applications on App Engine using JRuby, but because App Engine does not provide any support for relational databases, you'd need to heavily modify or completely rewrite your ActiveRecord code using DataMapper. But App Engine might be a good solution for your scaling problems because you get access to a high-efficiency, nonrelational data-storage system, and scaling for many users is almost automatic. You can download the Google App Engine SDK (http://code.google.com/appengine/downloads.html) if you want to try App Engine on your local development system.

PART 4

■ ■ ■

Information Publishing

Part 4 of this book deals with information publishing and large-scale distributed data processing to support information publishing. I will work with public data sources and create an example "mashup" web application, and then I will cover two options for large-scale data processing. Finally, I will close this book with one of my own projects: a web application for keeping track of interesting things.

CHAPTER 13

■ ■ ■

Creating Web Mashups

So far, we have mostly been working with techniques for back-end, server-side programming. This chapter is a bit of a departure, and it should be fun: you'll create a mashup web application using the Twitter web services and Google Maps web services to display locations mentioned in your friends' Twitter "tweets." Mashups use either public APIs from existing web sites and/or web scraping to collect data and combine it on a single web page.

I won't address security issues in this example. Any public web application that allows users to enter account information for other web sites needs to take suitable precautions safeguarding this information. Implementing security is more difficult when you're dealing with account names and passwords for third-party web portals because you cannot store cryptographic salts and password hashes; rather, you must store plain-text account information. So in this chapter, I am going to build a "personal mashup" that you can run on your laptop. It will require that you set Twitter and Google Maps account names and passwords in environment variables. Our example app will use these environment variables:

```
GOOGLEMAPS_API_KEY
TWITTER_ACCOUNT
TWITTER_PASSWD
```

Tip While keeping your passwords set in environment variables is fine for personal web-portal experiments, there are standards for open authentication APIs like OAuth (`http://oauth.net/`) that are better single sign-on (SSO) solutions for web sites requiring user authentication or login.

In the next two sections, we will take a look at the Twitter and Google Maps APIs and use two available gems to access these services. I'll then close out this chapter by developing a Rails web app that uses these third-party web services. All web-service calls to these web services are done on the server side in the Rails application. For third-party web services that do not require authentication, a very good alternative is to make the web-service calls from the client's browser using JavaScript.

■**Tip** Peter Szinek recommends HTTParty (`http://railstips.org/2008/7/29/it-s-an-httparty-and-everyone-is-invited`), which provides a common Ruby interface to a variety of REST web services.

Twitter Web APIs

Twitter documents its APIs at `http://apiwiki.twitter.com/`, and it is helpful to review the low-level APIs that Twitter supports. The Twitter APIs can return XML, JSON, RSS, and Atom payloads.

Twitter API Overview

Twitter uses a REST-type architectural style. You fetch data using HTTP `GET`, add data using `POST`, and delete data using `DELETE`. Standard HTTP status codes are used: 200 for OK, 401 for Unauthorized, and so on. Twitter supports OAuth (`http://oauth.net/`) authentication, which you should use if you publicly deploy a mashup like the example in this chapter. While some APIs do not require authentication (such as the API that lets you get the most recent public timelines for all of Twitter), I think that the interesting APIs are those that work with an authenticated user to provide results specific to that user's context. In the case of our mashup, this user will be you. As an example, once you are authenticated, getting the latest 20 "tweets" made by your friends is as easy as hitting this URL:

```
http://twitter.com/statuses/friends_timeline.xml
```

In this example, I requested XML data. You can find complete documentation for the REST APIs at `http://apiwiki.twitter.com/Twitter-API-Documentation`.

■**Tip** Twitter REST requests use a file extension (".xml" in the last example) to specify the requested return data type. I slightly prefer using HTTP request headers to specify return data types, and I cover this technique in Appendix B.

While making the web-service calls is simple, writing the code for authentication is a little more involved. So we will take an easier route and use the Twitter client gem written by John Nunemaker that handles authentication for you. You first need to install it: `gem install twitter`.

Using the Twitter Gem

The `twitter` gem adds a "Ruby-like" interface to the REST APIs by abstracting web-service calls as Ruby method calls and by converting returned payloads to Ruby objects. The `twitter` gem has a nice feature that I will not use in this chapter, but you can have fun with it on your own: the gem installs command-line tools for accessing Twitter and maintains local state in a

SQLite 3 database. The README file in the gem contains examples for using the command-line interface.

■Tip As usual, you can determine the location of Ruby gems for a specific Ruby installation using either of the two following commands on Linux, Unix, and Mac OS X systems: `gem environment` or `gem environment | grep INSTALLATION`.

Using the `twitter` gem is simple, especially if you are already familiar with the Twitter APIs. (The gem README file contains examples for search and timelines, but you won't need them in this case.) Later, we will use the following code snippets in the Rails web application. I start by getting my Twitter account information from my local environment-variable settings and loading the gem:

```
T_ACCOUNT=ENV['TWITTER_ACCOUNT']
T_PASSWD=ENV['TWITTER_PASSWD']
require 'rubygems'  # needed for Ruby 1.8.x
require 'twitter'
```

Now, you can create a Twitter `Base` class instance for your account with the following expression:

```
Twitter::Base.new(T_ACCOUNT, T_PASSWD)
```

My Twitter account is `mark_l_watson`, and I can get my complete user profile using:

```
Twitter::Base.new(T_ACCOUNT, T_PASSWD).user('mark_l_watson')
```

This returns data like the following (edited for brevity):

```
#<Twitter::User:0x71f430
 @description=
  "I am an author of 14 books (Java, artificial intelligence, Linux, etc.) and ➥
a Ruby/Java consultant",
 @followers_count="114",
 @friends_count="91",
 @location="Sedona Arizona",
 @name="mark_l_watson",
 @profile_image_url=
  "http://s3.amazonaws.com/twitter_production/profile_images/74285035/➥
Mark_hat_small_normal.jpg",
 @url="http://markwatson.com">
```

If you want to update your Twitter status (that is, what you are currently doing), use this code:

```
Twitter::Base.new(T_ACCOUNT, T_PASSWD).update("out hiking")
```

If you want to get the status of all your friends, you can use this:

```
Twitter::Base.new(T_ACCOUNT, T_PASSWD).friends.collect { |friend|
    [friend.name, friend.status.text] if friend.status
}.compact
```

This code fragment produces an array or arrays. Each inner array contains a friend's Twitter name and what he or she is currently doing. Here's some example output:

```
[["Tom Munnecke",
  "checking out my my video interview of Jon Haidt in Santa Barbara last week."],
 ["patrickdlogan",
  "sorry. I have chocolate in hand now and things are going better."],
 ["mattwagner",
  "It's nice when good news outshines bad news first thing in the morning."],
 ["jansaasman",
  "my son is now following me on Twitter. I hope he doesn't get disappointed."],
 ["Dion Almaer",
  "'-webkit-user-select: none' Thank you! Fixed the annoying 'selecting divs ➥
and objects' bug that we had."],
 ["Ben Goertzel",
  "Thinking about recording music specifically for parrots to listen to ... ➥
my daughter tried it last week but I think I could do better!"]]
```

Using the `twitter` gem makes it easy to interact with Twitter. We will plug these examples into our Rails mashup web application later, but first I will give you an overview of the Google Maps APIs.

Google Maps APIs

The Google Maps APIs enable you to place maps on web pages. You can specify a map's geographic range, and you can specify a map's center as latitude/longitude coordinates or as the name of a city or country. You need to register the domain of your web site where you will be using Google Maps. If you're running web applications on your laptop (as opposed to a server with a permanent IP address), you must sign up at http://code.google.com/apis/maps/signup.html for a key for the domain http://localhost:3000. From Google's FAQ: "If you are developing on localhost, you will need to register a key for http://localhost." For the examples in the rest of this chapter, I assume that you have an API key and that the value of the environment variable GOOGLEMAPS_API_KEY is set to your key value.

Google Maps API Overview

The most common use of the Google Maps APIs is placing JavaScript calls to the web service directly in web pages; I will provide an example that shows you how to do this with some manual coding. In the next section, you will see how to use a Rails plugin to initialize data from Google's web services and generate the JavaScript for your views. In addition to specifying map locations using latitude/longitude pairs, you can also use geolocation support that determines latitude and longitude from a place name like "San Francisco" or a street address.

You might want to refer to Google's online documentation at http://code.google.com/apis/maps/ after reading through the following short example. I derived it from Google's example at http://code.google.com/apis/maps/documentation/examples/map-simple.html and modified it to use geolocation; you can find it in src/part4/mashup_web_app/googlemap.html (get the code samples for this chapter in the Source Code/Download area of the Apress web site). Google recommends that you use strict XHTML to help prevent browser-compatibility problems. Here is the XHTML using geolocation to find the town where I live (Sedona, Arizona):

```
<!DOCTYPE html PUBLIC "-//W3C//DTD XHTML 1.0 Strict//EN"
    "http://www.w3.org/TR/xhtml1/DTD/xhtml1-strict.dtd">
<html xmlns="http://www.w3.org/1999/xhtml" xmlns:v="urn:schemas-microsoft-com:vml">
  <head>
    <meta http-equiv="content-type" content="text/html; charset=utf-8"/>
    <title>Google Maps: Using Geolocation to find Sedona Arizona</title>
    <!-- Load Google Maps Javascript and register key at the same time: -->
    <!-- Replace "KEY" with your own key value: -->
    <script src="http://maps.google.com/maps?file=api&amp;v=2.x&amp;key=KEY"
                type="text/javascript">
    </script>
    <script type="text/javascript">
        function initialize() {
          if (GBrowserIsCompatible()) {
            var map = new GMap2(document.getElementById("map_canvas"));
            var geocoder = new GClientGeocoder();
            geocoder.getLatLng(
              "Sedona",      <!-- Replace this value -->
              function(point) {
                  map.setCenter(point, 13);
              }
            );
            map.setUIToDefault();
          }
        }
    </script>
  </head>
  <!-- As per Google's documentation, call GUnload to free memory -->
  <body onload="initialize()" onunload="GUnload()">
    <div id="map_canvas" style="width: 500px; height: 300px"></div>
  </body>
</html>
```

You need to replace KEY on the ninth line of this listing with the API key that you generated. Figure 13-1 shows the map that's generated in a web browser after this example file is loaded.

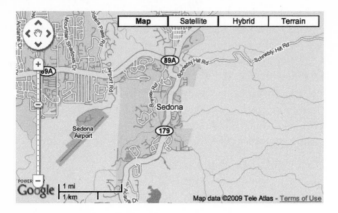

Figure 13-1. *Google Map generated from a static HTML file*

Using Google's online documentation and its XHTML "boilerplate," you can easily add maps to web pages. Refer to the documentation if you want to add overlays, custom controls, and so on. In the next section, you will see an easy way to add maps to Rails web applications by using the YM4R/GM Rails plugin.

Using the YM4R/GM Rails Plugin

I will be using both the YM4R/GM Rails plugin and the `twitter` gem in the next section, when we write our Rails mashup web application. I already have the YM4R/GM plugin installed in the example directory `src/part4/mashup_web_app`. When you need to install this plugin in new Rails applications, run the following script inside your top-level Rails directory:

```
ruby script/plugin install  ➥
svn://rubyforge.org/var/svn/ym4r/Plugins/GM/trunk/ym4r_gm
```

Running this script adds the plugin to the `vendor/plugins` directory. It also writes the file `config/gmaps_api_key.yml`, which you should edit to look like this:

```
development:
    ENV['GOOGLEMAPS_API_KEY']

test:
    ENV['GOOGLEMAPS_API_KEY']

production:
    YOUR_DOMAIN.com: ENV['GOOGLEMAPS_API_KEY']
```

In my `gmaps_api_key.yml` file, I get my API key from my local environment-variable settings instead of hard-coding it. But you can simply hard-code in your API key if you'd like.

Installing the plugin also installs several JavaScript files in `public/javascripts`: `geoRssOverlay.js`, `markerGroup.js`, `wms-gs.js`, and `ym4r-gm.js`.

The following listing shows a sample Rails view file that you could use to display maps. A view that displays maps needs to include the generated code from the plugin in the HTML header, and the generated map in the body:

```
<html>
  <head>
      <title>Demo From Plugin Documentation</title>
      <%= GMap.header %>
      <%= @map.to_html %>
  </head>
  <body>
        <%= @map.div(:width => 600, :height => 400) %>
  </body>
</html>
```

The variable @map is set in the controller for this view. An example controller method might be:

```
class SampleMapController < ApplicationController
  def index
      results = Geocoding::get("Sedona")
      @map = GMap.new("map_div")
      @map.control_init(:small_map => true,:map_type => true)
```

You need to check that the geolocation API call returned at least one result. If multiple results are returned, then use the first one for the map display:

```
      if results.length > 0
        @map.center_zoom_init([results[0].latitude,results[0].longitude],12)
        @map.overlay_init(GMarker.new([results[0].latitude,results[0].longitude],
                            :title => "test title",
                            :info_window => "test text for info window"))
      end
    end
  end
end
```

This example used the hard-coded location "Sedona" (the town where I live in the mountains of Central Arizona). In the next section, you'll use the Twitter APIs to get status updates for a list of friends, look for place names in the status messages, and use one of those place names to determine the map's display area.

An Example Rails Mashup Web Application

You now have everything you need for the example mashup web application: the code to find place names that you saw in Chapter 3, the twitter gem, and the YM4R/GM Rails plugin to create Google Maps. In this section, I will roll this all up into a simple web application (see Figure 13-2).

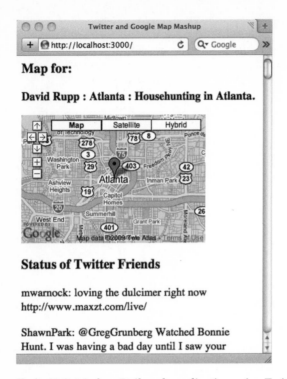

Figure 13-2. *Mashup Rails web application using Twitter and Google Maps*

Place-Name Library

As a reminder, you can refer to Chapter 3 for the implementation of the class TextResource. Also see the section "Extracting Entities from Text" in that chapter to review the implementation of the EntityExtraction class. In order to make this mashup example more lightweight, I extracted the 40 lines of code from the class TextResource that detects place names and I refactored this code into the file mashup_web_app/lib/Places.rb. I copied the data file placenames.txt to the directory mashup_web_app/db. The file Places.rb defines the module Places and the class EntityExtraction. The following expression will return an array of place names from input text:

```
Places::EntityExtraction.new('I went hiking in Sedona')
```

The output from evaluating this expression is:

```
["Sedona"]
```

I do not list the entire contents of Places.rb here because it is largely the same code that you saw in Chapter 3.

MashupController Class

Our example application will also need the Rails controller class MashupController, which uses both the twitter gem and the YM4R/GM Rails plugin. The index method on this class gets Twitter friends' status messages, finds place names within those messages, and shows a message containing a place name with a Google Map centered on that location. The following code snippets show the implementation of the method index, interspersed with explanations:

```
def index
```

The web-service call to get a long list of Twitter user-status objects is an expensive operation, so the Twitter documentation asks developers to cache the returned value and reuse it. But storing this data will cause an error: you'll exceed the 4K-byte limit imposed on storing user sessions in browser cookies. I'll address this problem in the next section, "Handling Large Cookies." In the meantime, I use this code to check whether I have cached the results of a Twitter web-service call:

```
if session[:hits]
  @hits, @all_hits = session[:hits]
else
```

The Twitter results have not been cached, so we make the web-service call:

```
friends_status =
  Twitter::Base.new(T_ACCOUNT, T_PASSWD).friends.collect { |friend|
    [friend.name, friend.status.text] if friend.status
  }.compact
@hits = []
@all_hits = []
```

The variable @hits contains only status updates that contain place names, and the variable @all_hits contains all status updates:

```
friends_status.each {|name, status|
  @all_hits << [name, status]
  Places::EntityExtraction.new(status).place_names.each {|place_name|
    @hits << [name, place_name, status]
  }
}
end
```

Next, we randomly choose a status update that contains a place name and use the YM4R/GM plugin to set up data for a map:

```
@chosen_hit = @hits[rand(@hits.length)]
if @chosen_hit
```

The following expression uses a place name to calculate the latitude/longitude coordinates for the place:

```
results = Geocoding::get(@chosen_hit[1])
```

The variable @map will be used in the view for this controller method:

```
@map = GMap.new("map_div")
@map.control_init(:small_map => true,:map_type => true)
if results.length > 0
  @map.center_zoom_init([results[0].latitude,results[0].longitude],12)
  @map.overlay_init(GMarker.new([results[0].latitude,results[0].longitude],
                                 :title => "#{@chosen_hit[0]}: ➡
                                 #{@chosen_hit[2]}",
                                 :info_window => "Info! Info!"))
  #@map.overlay_init(GMarker.new(@chosen_hit[1],:info_window => @chosen_hit[0]))
  end
 end
end
```

Handling Large Cookies

The controller shown in the last section stores a lot of data in a user's session cookie. This is fine because our example web application is designed as a single-user system. But as I mentioned earlier, the default user-session storage for Rails is a cookie that is limited to 4K bytes, which will present a problem if you want to scale out the application. You can work around this limitation by storing session data in a database. To do this, you must run two rake tasks to create migration files and create the session-data table:

```
rake db:sessions:create
rake db:migrate
```

Now session data will be stored in the default SQLite 3 database that I configured when I initially created this web application. In order to use this new session database, you need to add the following line at the end of the file config/initializers/session_store.rb:

```
ActionController::Base.session_store = :active_record_store
```

Now session data will be stored in a server-side database instead of the user's web browser.

■**Tip** In general, user sessions should not contain very much data, and the 4K-byte limit for each user's session data makes a lot of sense. You want to be especially careful to minimize the size of session data for web applications that serve many simultaneous users. Rails web applications are supposed to scale using a "shared nothing" approach, where, for example, incoming requests can be sent to any Mongrel process on any server running Mongrels for a web application. What is not "shared nothing" is persistent storage of user sessions, and sometimes one of the first scaling bottlenecks encountered in horizontally scaling an application is the single-user session data store for a web application. Making session data smaller will help postpone this particular scaling problem. If performance is an issue, consider storing session data in memcached.

Rails View for a Map

The view for showing the map and the Twitter status messages uses generated header data from the YM4R/GM plugin and data in the variable @map that was assigned a value in the MashupController class's index method:

```
<html>
  <head>
    <title>Twitter and Google Map Mashup</title>
    <%= GMap.header %>
    <%= @map.to_html %>
  </head>
  <body>
```

Here we show a randomly chosen status message that contains a place name with a Google Map showing the location of the selected place. The variables @chosen_hit and @map were set in the controller:

```
    <h3>Map for:</h3>
    <p><strong><%= @chosen_hit.join(' : ') %></strong></p>
    <%= @map.div(:width => 300, :height => 200) %>
    <h3>Status of Twitter Friends</h3>
```

The following embedded Ruby code shows all recent status messages for friends. The variable @all_hits was set to an array of arrays. The inner arrays contain two strings, a person's account name and his or her status message:

```
    <% @all_hits.each begin |hit| %>
        <p><%= hit.join(': ') %></p>
    <% end %>
  </body>
</html>
```

Output generated by this view is shown in Figure 13-2.

Wrapup

Mashup applications are becoming more prevalent on the Web, as you've probably noticed. They mostly use popular web services such as Google Maps, Flickr, Twitter, Facebook, and so on. I expect that the idea of mashups will extend to building web interfaces that combine information both from private web services and from niche, vertical-market web services.

Web services are versatile in that both humans and other software systems can use them. In programming, the Don't Repeat Yourself (DRY) philosophy specifies that you should not have duplicated code that does essentially the same thing. For Web 3.0 development, the DRY principle guides us to build intelligent, semantically tagged data sources just once and to reuse them for various human-facing devices and other software services. Web service–based data sources become reusable components.

If a company invests in building web services to provide customer data, then it should make additional smaller investments to write clients for its sales staff's cell phones, web browsers, end-of-year reporting applications, and so on. If other web services are developed to track sales and in-house staffing data, then it certainly makes sense to create mashup applications to integrate and present information. Except for issues with user authentication and single sign-on, writing mashup applications will probably require little work compared to developing and maintaining the back-end web services.

CHAPTER 14

■ ■ ■

Performing Large-Scale Data Processing

Some Web 3.0 applications will need to process very large data sets. In some cases you can use vertical scaling (with a single server), in which case you can fit entire data sets in memory or process data in segments that you can swap in from disk storage. The phrase "vertical scaling" usually refers to scaling by upgrading servers with more memory, CPU cores, and so on; I won't cover it in this chapter, though. Instead, I'll cover horizontal scaling that assumes multiple available servers, but you don't need to max out the individual servers with memory and other resources. Specifically, we are going to use the distributed map/reduce algorithm as implemented in the Hadoop open source project (`http://hadoop.apache.org`). Hadoop is based on Google's distributed map/reduce system.

I am going to give you a quick introduction to Hadoop by covering some theory and setting up a local Hadoop installation. Then I'll work through examples using some of the text-mining algorithms that we used in Chapter 3, but in a way that can scale to very large data sets. Specifically, I'll show you how to create streaming map/reduce applications that serve as inverted indices for search engines: you'll build an inverted word index and an inverted person-name index using Ruby, and then a similar inverted person-name index using Java. We can do all of this using a Hadoop system running on a single server in development mode. If you want to run Hadoop on your own servers, I will refer you to documentation on the Apache Hadoop web site for setting up a cluster of Hadoop servers. I will close out this chapter by showing you how to use Amazon's Hadoop web services. Once you design, develop, and debug map/reduce functions on your local Hadoop development system, you can run your functions on Amazon's system.

Hadoop is written in Java, which is probably the best language for writing Hadoop map/reduce plugins. Java map/reduce application code can take full advantage of direct interaction with the Hadoop system. However, Hadoop also supports streaming applications that take input from standard input and write their output to standard output. You can write streaming map/reduce application code in any programming language; I will use Ruby and Java in my streaming examples. If you are not a Java developer, you can skip the short section with the Java example. However, you have direct access to Hadoop APIs when writing map/reduce applications in Java.

Hadoop relies on the Hadoop Distributed File System (HDFS). For our purposes, I will treat HDFS as a black box because it is included in the Hadoop distribution. The Hadoop installation process also initializes HDFS. At the end of the chapter, when we use the Amazon

Elastic MapReduce web service, both Hadoop and HDFS will be all set up for us. I am going to cover map/reduce at a higher level in the next section, after which we will implement three examples of map/reduce applications.

Using the Distributed Map/Reduce Algorithm

Map/reduce works with data that can be expressed as key/value pairs. For our purposes in this chapter, we assume that data is streamed between application components that we supply to Hadoop. Each key/value pair is assumed to appear on one line of data (separated with a new-line character) in the incoming and outgoing data streams. Because you supply both the input data and the map and reduce functions, you get to decide how to interpret data on each incoming line. If key values are strings that might contain multiple words, it makes sense to use a tab character to separate the key data and the value data in each line. Figure 14-1 shows the initial map phase that involves parsing input documents and writing key/value pairs consisting of a word and a document index out to intermediate files. Figure 14-1 shows just one input data segment. Typically, input data is divided into many smaller segments and the segments are assigned to free map processes.

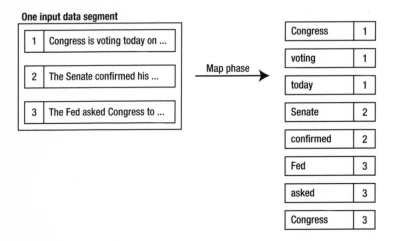

Figure 14-1. *Input text documents identified by unique IDs (1, 2, and 3) are split into segments for processing by multiple map processes.*

Map functions are usually described as performing some operation on input data streams and writing new key/value pairs to an output stream. The reason the map/reduce algorithm is distributed is that we can have multiple copies of both the map and reduce functions running in parallel, possibly on a very large number of servers. You might think that you need to do extra software development work to make your application-specific map and reduce functions manage execution in a distributed environment—but you don't! The trick is that the Hadoop system sorts by key values the output streams from each mapping function and effectively merges the output from multiple map functions into a single sorted data stream. This same sorting process is repeated for the output stream generated from multiple copies of your application-specific reduce function. Figure 14-2 shows how the key/value pairs from the map phase are sorted by key values, segmented into smaller sets, and assigned to reduce processes.

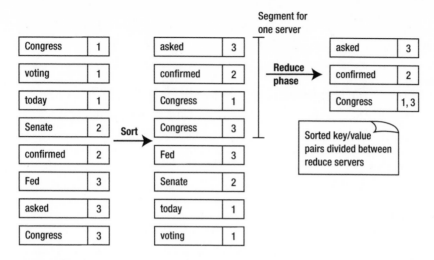

Figure 14-2. *Key/value pairs are sorted and divided into small segments for processing by multiple reduce processes. The reduce phase creates an inverted word index in this example.*

At the small cost of "reprogramming" yourself, you will quickly get used to solving problems using map/reduce in this chapter. Before you start to enjoy writing applications based on map/reduce functions, you'll do some setup work in the next section by installing Hadoop.

Installing Hadoop

The Hadoop web-site documentation goes into detail about setting up Hadoop, but I will give you some brief instructions here. They should help you set up Hadoop quickly so you can start developing map/reduce applications on your laptop.

■Note If you run Hadoop on a single server or on your laptop for development, the results should be identical to running the same problem on a cluster of Hadoop servers.

I am using the latest stable version of Hadoop (hadoop-0.18.3) for these examples. Later development versions are slightly incompatible, so you might want to start by using the same version I am and wait until later to try the next stable version.

I edited the file bin/hadoop-config.sh by adding this line near the top of the file:

```
export JAVA_HOME=/System/Library/Frameworks/JavaVM.framework/Versions/1.6/Home
```

You need to set this environment variable to the directory on your system that contains your Java development kit. Even though we are working in development mode on a single server (I am using a MacBook), setting JAVA_HOME in your environment is probably not adequate because all Hadoop commands are run through the Secure Shell (SSH) protocol and settings in your .profile file will not be picked up. Setting this environment variable in your .bashrc file is fine, but editing the hadoop-config.sh file is more straightforward.

You also need to perform one-time initialization using the following command:

```
bin/hadoop namenode -format
```

On my system, this one-time setup command writes the file system to:

```
/tmp/hadoop-markw/dfs/name
```

If you are running the Hadoop server in production mode, the following command starts all configured instances:

```
bin/start-all.sh
```

However, for the purposes of developing application-specific map and reduce functions, I will show you later how to quickly start up your own Ruby or Java applications each time you want to make a test or debug run (see the section "Running the Ruby Map/Reduce Functions").

Writing Map/Reduce Functions Using Hadoop Streaming

You can write streaming map/reduce functions in any language. Hadoop streaming (http://wiki.apache.org/hadoop/HadoopStreaming) uses the Hadoop system to control map/reduce data-processing jobs. You supply shell scripts for the map and reduce functions. These shell scripts are automatically copied to servers running Hadoop map and reduce operations for your data-processing jobs. If you invoke external programs in these scripts, then these programs and any external data that they require must be installed on all the servers in your Hadoop cluster.

Before developing our three streaming map/reduce applications, I will first show you in the next section how to run Ruby map and reduce scripts.

Running the Ruby Map/Reduce Functions

The commands for running Ruby map/reduce jobs and Java map/reduce jobs are different. (I will show you a Java example later in this chapter, in the section "Creating an Inverted Person-Name Index with Java Map/Reduce Functions.") In development mode, I place all of my input text in the input subdirectory of my Hadoop installation directory. When Hadoop runs, all files in the input subdirectory are processed. Before running a map/reduce job, you need to delete the output subdirectory; Hadoop will stop with an error if you try to skip this step.

Assuming that you want to run one of the examples that I write in the next section, and that the scripts are named map.rb and reduce.rb, you can start a job in development mode using this command:

```
bin/hadoop jar contrib/streaming/hadoop-0.18.3-streaming.jar        ➥
                     -mapper map.rb -reducer reduce.rb              ➥
                     -input input/* -output output                 ➥
                     -file /Users/markw/Documents/WORK/hadoop-0.18.3/map.rb    ➥
                     -file /Users/markw/Documents/WORK/hadoop-0.18.3/reduce.rb
```

You can find all of the Ruby map/reduce example scripts that I wrote for this chapter in the ZIP file Ruby_map_reduce_scripts.zip (you can get the code samples for this chapter from the Source Code/Download area of the Apress web site). This ZIP file contains:

```
 Length     Date    Time    Name
--------    ----    ----    ----
 2852824  04-13-09  15:51   peoplemap.rb
     593  04-13-09  15:52   peoplereduce.rb
     148  04-11-09  16:32   map.rb
     494  04-13-09  14:05   reduce.rb
```

We'll work with the Ruby scripts map.rb and reduce.rb in the next section to create an inverted word index that you can use to build a search engine. In the section after that, we will also use the peoplemap.rb and peoplereduce.rb scripts to build an inverted index of people's names found in the input data set. The file peoplemap.rb is 2.8 megabytes long because I encoded name-lookup hash tables in the script file.

Creating an Inverted Word Index with the Ruby Map/Reduce Functions

Building an inverted word index is the first step in providing a search service like Yahoo! or Google. If you have a fairly small set of input documents or web pages to index, then it is simple to write a short Ruby script that uses RAM for working space to create inverted word indices. The approach that I use in this chapter is very different because you can run it on a large number of servers and it uses disk for intermediate and working storage.

■Note The Hadoop system is modeled after the Google File System (GFS) and Google's distributed map/reduce infrastructure. Yahoo! uses Hadoop to implement some of its search services.

You can segment your input data into any number of individual files. I created two short input test files, which you can find in the Hadoop input directory. These files contain very little data, so it will be easy for you to look at intermediate results of a map/reduce run and better understand the distributed map/reduce algorithm. The file test1.txt contains:

```
doc3 John Smith and Bob Jones
doc4 John Smith and Bob Smith
```

The file test2.txt contains:

```
doc1 The cat ran up the street
doc2 Bob Brown went fishing
```

As I have already mentioned, because we're writing our own map and reduce functions, we are free to interpret the data on each input line in any way we need to for any specific application. In this example, we interpret each line in the input stream as a single input document. The line begins with the document name, which is terminated by a space character. Everything after the first space character is interpreted as the words in the document represented by that input line.

If you were indexing web pages, you would extract the plain text from each web page. You could treat the URL for the web page as the document name. For each web page that you wanted to index, you would write the document name (web-page URI) followed by a space character and all of the text on the web page. Because Hadoop uses new-line characters to separate each input line sent to your map function, you must remember to strip all new-line characters from the plain text that you scrape from a web page. I will show you a Ruby script in the section "Running with Larger Data Sets" that I used to scrape Wikipedia article data and write it to Hadoop input files.

The script map.rb contains a map function that processes all lines in an input stream. Each line is split into words, and the first word is taken to be the name of a source document. Each word after the document name is written to the output stream with the document name. I am using a tab character to separate a word from the name of the document it is contained in. Here is a listing showing the file map.rb:

```ruby
#!/usr/bin/env ruby

STDIN.each_line do |line|
  words = line.split
  doc = words.shift
  words.each do |word|
    puts "#{word}\t#{doc}"
  end
end
```

Notice that for this inverted word index application, I am reversing the order from the input stream: now the word is the key and the document name is the value. The output stream that the map.rb script generates and sends to the reduce function looks like this:

```
Bob       doc2
Bob       doc3
Bob       doc4
The       doc1
and       doc3
and       doc4
cat       doc1
fishing       doc2
```

```
ran     doc1
street      doc1
```

You need to understand that many copies of your map function can carry out this map processing, and that you can run these multiple copies at the same time on different servers. Hadoop will split up the data from your input files and send chunks of input text to the multiple map functions. The output streams from all the map operations are merged together into one output stream that is sorted by key value. Hadoop does not know how much of each input line your application uses as a key value, so the entire line can be used for the sorting process. This sorted output stream from the combined map runs is partitioned into chunks and sent off to multiple copies of your reduce function.

The reduce.rb script for this example assumes that the input stream is sorted by words (the keys), so each copy of the reduce function can collect multiple documents for a word and write the document list along with the word as a single line in the output stream. To make this really clear, I'll manually follow the process for the case where my reduce function gets a stream with four lines:

```
Bob     doc2
Bob     doc3
Bob     doc4
The     doc1
```

We want to record the fact that the word "Bob" occurs in the documents doc2, doc3, and doc4. We would want to write a single line to the output stream:

```
Bob     doc2 doc3 doc4
```

You might think that the combined output stream from the map functions could be partitioned into two chunks, each containing the key value "Bob." In this case, we would have a problem: the final output from combining the output streams of all *reduce* functions could contain two lines that started with the key value "Bob." Fortunately, Hadoop works around this possibility by trying to break data streams into chunks where key values change, so all lines with the same key get sent to a single map or reduce function. If this is not possible, then you can use an additional reduce phase to coalesce all indices for a given key. In general, the number of intermediate data segments far exceeds the number of processes running map and/or reduce functions.

In my reduce function in the following listing, I use a temporary array to store the names of all documents containing a given key value. When the key value in the input stream changes, I write the previous key value with the stored document names to the output stream:

```ruby
#!/usr/bin/env ruby

last_doc, last_word = '', ''
doc_list = []

STDIN.each_line do |line|
  word, doc = line.split("\t").collect {|w| w.strip}
```

```
  STDERR.puts("#{word} : #{doc}")
  if last_doc != '' && last_word != word
    STDERR.puts("doc_list: #{doc_list}")
    puts "#{last_word}\t#{doc_list.join(' ')}"
    doc_list = [doc]
    last_word = word
    last_doc = doc
  else
    doc_list << doc.strip
    last_word = word
    last_doc = doc
  end
end
puts "#{last_word}\t#{doc_list.join(' ')}"
```

Just as in the map function, you write to standard output using the method `puts` to write to the output stream. If I write to standard error (`STDERR.puts("doc_list: #{doc_list}")` in this example), then whatever information I print appears in the output from Hadoop when you are running in development mode. In order to effectively write debug output to standard error, you should initially test your map and reduce functions using small input data sets.

The sorted and merged output from all reduce tasks is written to the subdirectory `output`. Because this is a very small test with little input, the output is written to the single file `output/part-00000`, which looks like this:

```
Bob      doc2 doc3 doc4
John     doc3 doc4
Smith    doc3 doc4 doc4
and      doc3 doc4
fishing  doc2
street   doc1
up       doc1
```

■**Note** Remember that Hadoop always sorts the output from the map and reduce functions by key value before sending it to the next processing stage.

In the next section, I will show you another map/reduce application that creates an inverted index of people's names found in input text.

Creating an Inverted Person-Name Index with the Ruby Map/Reduce Functions

The implementation of this example is similar to the inverted word index example in the last section. The primary difference is that instead of just tokenizing words from input text,

I extract human names and discard all other input text. I will use this inverted person-name index in a Rails web application example in Chapter 15.

The map function defined in the file `peoplemap.rb` contains code that locates human names in input text, similar to what you saw in Chapter 3. This file also contains more than 100,000 lines of data-initialization code because I wanted this Ruby script to be entirely self-contained. When you're distributing map/reduce functions across multiple servers, any libraries or static data sets that are required in your functions must be available in the runtime environments on those servers. So you should make your Ruby map and reduce scripts self-contained when possible. I also make the Java examples in the next section self-contained for the same reason.

As I mentioned, `peoplemap.rb` contains about 100,000 lines of data-initialization code. Here are a few of them:

```
FIRST_NAMES['Reggie']=true
FIRST_NAMES['Rena']=true
FIRST_NAMES['Chaz']=true
FIRST_NAMES['Morgan']=true

LAST_NAMES['Hoyos']=true
LAST_NAMES['Bruffee']=true
LAST_NAMES['Iguina']=true
LAST_NAMES['Bellus']=true
LAST_NAMES['Nowotka']=true
LAST_NAMES['Kolkhorst']=true

PREFIX_NAMES['Vice']=true
PREFIX_NAMES['Lt']=true
PREFIX_NAMES['Senator']=true
```

The top-level method `get_names` looks for names in an array of input words and returns an array of human names (this is much the same code from the "Extracting Entities from Text" section in Chapter 3, but I simplified it to extract only human names rather than human names plus place names):

```
def get_names words
  word_flags = []
  words.each_with_index  {|word, i|
    word_flags[i] = []
    word_flags[i] << :first_name  if FIRST_NAMES[word]
    word_flags[i] << :last_name    if LAST_NAMES[word]
    word_flags[i] << :prefix_name if PREFIX_NAMES[word]
  }

  # easier logic with two empty arrays at end of word flags:
  word_flags << [] << []

  # remove :last_name if also :first_name and :last_name token nearby:
  word_flags.each_with_index  {|flags, i|
    if flags.index(:first_name) && flags.index(:last_name)
```

```
      if word_flags[i+1].index(:last_name) || word_flags[i+2].index(:last_name)
        word_flags[i] -= [:last_name]
      end
    end
  }

  # look for middle initials in names:
  words.each_with_index {|word, i|
    if word.length == 1 && word >= 'A' && word <= 'Z'
      if word_flags[i-1].index(:first_name) && word_flags[i+1].index(:last_name)
        word_flags[i] << :middle_initial if word_flags[i].empty?
      end
    end
  }

  # discard all but :prefix_name if followed by a name:
  word_flags.each_with_index  {|flags, i|
    if flags.index(:prefix_name)
      word_flags[i] = [:prefix_name] if human_name_symbol_in_list?(word_flags[i+1])
    end
  }

  #discard two last name tokens in a row if the preceding token is not a name token:
  word_flags.each_with_index  {|flags, i|
    if i<word_flags.length-2 && !human_name_symbol_in_list?(flags) && ➥
word_flags[i+1].index(:last_name) && word_flags[i+2].index(:last_name)
      word_flags[i+1] -= [:last_name]
    end
  }

  # discard singleton name flags (with no name flags on either side):
  word_flags.each_with_index  {|flags, i|
    if human_name_symbol_in_list?(flags)
      unless human_name_symbol_in_list?(word_flags[i+1]) || ➥
human_name_symbol_in_list?(word_flags[i-1])
        [:prefix_name, :first_name, :last_name].each {|name_symbol|
          word_flags[i] -= [name_symbol]
        }
      end
    end
  }

  human_names = []
  human_name_buffer = []
  in_human_name = false
  word_flags.each_with_index  {|flags, i|
    human_name_symbol_in_list?(flags) ? in_human_name = true : in_human_name = false
```

```
    if in_human_name
      human_name_buffer << words[i]
    elsif !human_name_buffer.empty?
      human_names << human_name_buffer.join(' ')
      human_name_buffer = []
    end
  }
  human_names.uniq
end

def human_name_symbol_in_list? a_symbol_list
  a_symbol_list.each {|a_symbol|
    return true if [:prefix_name, :first_name, :middle_initial, ➥
:last_name].index(a_symbol)
  }
  false
end
```

After the 100,000 lines of data and the top-level get_names method are defined, I place the map function at the end of the file peoplereduce.rb:

```
STDIN.each_line do |line|
  words = line.split
  doc = words.shift
  people = get_names(words)
  people.each do |person|
    #STDERR.puts("people_map.rb:  #{person} - #{doc}")
    puts "#{person}\t#{doc}"
  end
end
```

If you use the same input files we used in the last section, output written as input for the map function would look like the following line, where a tab character separates the human name from the name of the document containing it:

John Smith	doc2
John Smith	doc3

The reduce function needs to maintain local state to remember the documents for a specific key value. In this application, key values are people's names. When the key value changes in the input stream, the reduce function writes out the previous key value followed by a list of document names containing that key value:

```
#!/usr/bin/env ruby

last_doc, last_person_name = '', ''
doc_list = []
```

```ruby
STDIN.each_line do |line|
  person_name, doc = line.split("\t").collect {|w| w.strip}
  #STDERR.puts("#{person_name} : #{doc}")
  if last_doc != '' && last_person_name != person_name
    #STDERR.puts("doc_list: #{doc_list} #{last_person_name}")
    puts "#{last_person_name}\t#{doc_list.join(' ')}"
    doc_list = [doc]
    last_person_name = person_name
    last_doc = doc
  else
    doc_list << doc.strip
    last_person_name = person_name
    last_doc = doc
  end
end
puts "#{last_person_name}\t#{doc_list.join(' ')}"
```

If you use the same input files we used in the last section, output written to the output stream would look like the following line, where a tab character separates the human name from the *list* of documents containing the name:

```
John Smith doc3 doc4
```

Assuming that you have copied the two Ruby scripts `peoplemap.rb` and `peoplereduce.rb` to your Hadoop installation directory, copied input files to the subdirectory `input`, and deleted the subdirectory `output`, you can then run this example in development mode using the following command:

```
bin/hadoop jar contrib/streaming/hadoop-0.18.3-streaming.jar          ➡
          -mapper peoplemap.rb -reducer peoplereduce.rb               ➡
          -input input/* -output output                               ➡
          -file /Users/markw/Documents/WORK/hadoop-0.18.3/peoplemap.rb  ➡
          -file /Users/markw/Documents/WORK/hadoop-0.18.3/peoplereduce.rb
```

The output file is:

```
$ cat output/part-00000
Bob Brown    doc2 doc3
Bob Jones    doc3
Bob Smith    doc4
John Smith   doc3 doc4
```

If you want to run this example using a cluster of Hadoop servers, reference the documentation on the Hadoop web site (`http://hadoop.apache.org/core/docs/r0.18.3/cluster_setup.html`). I will show you later in this chapter how to run this example using Amazon's Elastic

MapReduce Hadoop-cluster web services (see the section "Running the Ruby Map/Reduce Example Using Amazon Elastic MapReduce").

Creating an Inverted Person-Name Index with Java Map/Reduce Functions

Now I'll show you how to use Java map/reduce functions to write an inverted person-name index that helps you find people's names in a large document collection. If you are not a Java developer, you can skip this example. The advantage of using Java is that you have full access to the Hadoop configuration and control APIs in your map/reduce applications. That said, the example in this section resembles the streaming-mode Ruby examples you have just seen.

You can find the Java example packaged up in a JAR file with this book's source code, ready to use. You can copy src/part4/namefinder.jar to your Hadoop installation directory. This JAR file contains compiled code, required data, and Java source code:

```
$ jar tvf namefinder.jar
        0 META-INF/
       60 META-INF/MANIFEST.MF
        0 com/
        0 com/knowledgebooks/
        0 com/knowledgebooks/mapreduce/
     3141 com/knowledgebooks/mapreduce/NameFinder$MapClass.class
     3015 com/knowledgebooks/mapreduce/NameFinder$Reduce.class
     3466 com/knowledgebooks/mapreduce/NameFinder.class
     6158 com/knowledgebooks/mapreduce/NameFinder.java
        0 com/knowledgebooks/nlp/
     4724 com/knowledgebooks/nlp/ExtractNames.class
    10066 com/knowledgebooks/nlp/ExtractNames.java
        0 com/knowledgebooks/nlp/util/
     5119 com/knowledgebooks/nlp/util/ScoredList.java
     3321 com/knowledgebooks/nlp/util/Tokenizer.class
     4547 com/knowledgebooks/nlp/util/Tokenizer.java
        0 data/
  1576682 data/peoplenames.ser
```

I copied the implementation pattern in the Hadoop Ruby examples by placing the static Map and Reduce classes inside a single source file, in this case NameFinder.java. The class com.knowledgebooks.nlp.ExtractNames is similar to the Ruby code in Chapter 3 and in the preceding section. You can read through the source code in the JAR file to see the Java implementation.

The map function is defined in the static class NameFinder$MapClass, which I'll show in the next code snippet. The generic org.apache.hadoop.mapred.Mapper interface has four arguments: input key, input value, output key, and output value 2. The keys and values must be instances of any class that is derived from the Hadoop class Writeable, which is the base class for key and value types that can be text strings, integers, long integers, and so on. In this example, all input and output keys and values are text strings:

```
public static class MapClass extends MapReduceBase
        implements Mapper<LongWritable, Text, Text, Text> {

    private Text human_name = new Text();
    private Text doc = new Text();

    public void map(LongWritable key, Text value,
                            OutputCollector<Text, Text> output,
                            Reporter reporter) throws IOException {
        String line = value.toString();
        //System.err.println("NameFInder: map: key="+key+" line="+line);
        int index = line.indexOf(" ");
        if (index > -1) {
            String doc2 = line.substring(0, index);
```

The argument key is a character index from the beginning of the input stream; you don't need this argument, so just ignore it. As with the streaming Ruby examples, if I write debug output to standard error, I see the printouts when running my map/reduce applications in development mode and see these printouts in my output logs if I am running on my own Hadoop cluster or on Amazon's Elastic MapReduce system:

```
             //System.err.println("NameFInder: map: doc2="+doc2);
            doc.set(doc2);
            List<String> names =
                    extractNames.getProperNames(line.substring(index));
            for (String name : names) {
                human_name.set(name);
                output.collect(human_name, doc);
            }
        }
    }
}
```

The output would be the same as in the Ruby example in the last section:

```
John Smith    doc3
John Smith    doc4
```

The reduce function is defined in the static class NameFinder$Reduce:

```
public static class Reduce extends MapReduceBase
        implements Reducer<Text, Text, Text, Text> {
    private String last_doc = "";
    private String last_person_name = "";
    private List<String> doc_list = new ArrayList<String>();
```

The interface for reduce functions differs from what you saw in the Ruby examples. Here, the method reduce is called for each unique key value with an iterator for all the values for that key:

```
public static class Reduce extends MapReduceBase
        implements Reducer<Text, Text, Text, Text> {

    public void reduce(Text person_name, Iterator<Text> documents,
                       OutputCollector<Text, Text> output,
                       Reporter reporter) throws IOException {
        String person = person_name.toString();
        List<String> doc_list = new ArrayList<String>();
        while (documents.hasNext()) {
            Text document = documents.next();
            String document_str = document.toString();
            if (document_str.substring(0, 1).equals("["))
                document_str = document_str.substring(1, document_str.length() - 1);
            doc_list.add(document_str);
        }
        output.collect(new Text(person), new Text(doc_list.toString()));
    }
}
```

You might have noticed that the logic is much simpler than the Ruby version because all values for a given input key value are passed in one call to the reduce method. In the Ruby streaming examples, the reduce function gets called with key/value input pairs, so I had to keep track of when a key value was changing in the input stream. I am not covering all the details of writing Java map/reduce applications here; check out the Hadoop documentation (http://hadoop.apache.org/core/docs/current/mapred_tutorial.html) for more details.

To run this example in development mode, copy your input files to the subdirectory input, delete the subdirectory output, and then run the following command:

```
bin/hadoop jar namefinder.jar com.knowledgebooks.mapreduce.NameFinder  ➥
                       -m 1 -r 1 input/ output/
```

You can use the command-line arguments -m and -r to control the number of map and reduce processes for a data run. The output file from the final reduce operation is written to the subdirectory output, and it looks like this:

```
$ cat output/part-00000
Bob Brown    [doc2, doc3]
Bob Jones    [doc3]
Bob Smith    [doc4]
John Smith   [doc3, doc4]
```

So far in this chapter, I have been using very small input data files to show you how map/reduce works and how to write data-processing applications. In the next section, I will build a large input data set.

Running with Larger Data Sets

We have been using a trivially small input data set for all our Hadoop examples so far. This worked well for my inverted word index and my inverted person-name index, the latter of which I implemented in both Ruby and Java.

In order to get a much larger and more useful input data set, I wrote a simple spider to fetch random Wikipedia articles. The Wikipedia license allows fairly free use of the material on Wikipedia, so it is a great source of text data. In all, I collected about 6,000 random articles, which you can find in the file src/part4/wikipedia_article_data.zip. I urge you to reuse the data that I collected rather than run the spider that I'll show you in this section, in order to save a load on the Wikipedia servers.

■**Caution** It is rude behavior to make many requests of public web servers in a short time period. Over a period of a few days, I collected the data by running the spider mostly at night when the Wikipedia servers would be lightly used, and I set a 15-second waiting period between requests for random web pages.

The output-text file format consists of this information on each line: the URL of a Wikipedia page, a tab character, then all of the plain text extracted from the article page. It is important that you remove all new-line characters from the plain text extracted from a Wikipedia article web page.

The following script, which resides in the file part4/get_wikipedia_data.rb, contains the top-level methods get_article_as_1_line and get_random_article:

```ruby
require 'rubygems' # only needed for Ruby 1.8.6
require 'open-uri'

def get_article_as_1_line uri
  ret = ""
  begin
    open(uri) { |inp|
      ret += "#{inp.base_uri}\t"
      inp.each do |line|
        ret << line.gsub(/<\/?[^>]*>/, " ")
        ret << " "
      end
    }
  rescue
    puts "Error: #{$!}"
  end
  ret.gsub("\n", "").gsub('\t',' ')
end

def get_random_article
  get_article_as_1_line "http://en.wikipedia.org/wiki/Special:Random"
end
```

Now that I have defined the two utility methods I need for fetching and preparing data for input to Hadoop, I can easily write several data files that can be copied to the Hadoop subdirectory `input`:

```
4.times {|iter|
  File.open("wikipedia_data_#{iter+1}.txt", 'w') do |f|
    1500.times {|i|
      f.puts get_random_article
      sleep(15)
    }
  end
}
```

Assuming that you have set up Hadoop on your laptop in development mode or followed the Hadoop documentation to set up a server cluster, I leave it as an exercise for you to copy the text files created by the preceding code example to your Hadoop `input` subdirectory and run either of the Ruby applications or the Java application.

I will now show you how to copy these text data files to Amazon S3, so we can use the files later with Amazon Elastic MapReduce. I use the open source `s3cmd` command-line tools (see `http://s3tools.org/s3cmd`) to add data to S3 and copy S3 data back to my own servers or laptop. For the rest of this chapter, I assume that you have installed the `s3cmd` tools. You also need an S3 account on Amazon if you want to follow along with the examples here and in the next section. You can use the sign-up links on the AWS Management Console (`https://console.aws.amazon.com/`) to get accounts to use both Amazon EC2 and Elastic MapReduce.

I used the following command to create a new bucket for the Wikipedia data:

```
s3cmd mb s3://web3-book-wikipedia
```

■**Note** Every bucket name on S3 must be unique, and the bucket names are not specific to your account. For example, because I created a bucket named `web3-book-wikipedia`, this name is taken and no one else can reuse it unless I delete this bucket.

I then copied the Wikipedia text data files to S3:

```
s3cmd put wikipedia_data_1.txt s3://web3-book-wikipedia/
s3cmd put wikipedia_data_2.txt s3://web3-book-wikipedia/
```

If you have an S3 account, you can copy the Wikipedia data from the `src/part4/wikipedia_article_data.zip` ZIP file. Using S3 storage is inexpensive, but not free. You can use the "Your Account" link on the AWS Management Console to monitor any charges that you might be accruing. A typical Elastic MapReduce run costs me about $0.15 to run, plus an additional charge for storing the input and output data on S3.

I will now show you how to run Hadoop jobs on Elastic MapReduce using this Wikipedia data.

Running the Ruby Map/Reduce Example Using Amazon Elastic MapReduce

You can set up Hadoop yourself either on your own servers or on Amazon EC2, which is a good approach if customizing your setup is important. However, a much easier way to use Hadoop is through the Amazon Elastic MapReduce system (http://aws.amazon.com/elasticmapreduce/). If you've written your map/reduce functions in Java and created a JAR file containing them, you can set up the entire process of running Hadoop map/reduce data-processing jobs with a web interface. Similarly, you can easily run both of the Ruby map/reduce application examples in this chapter using the Elastic MapReduce web interface. For this section, I will run the second Ruby application to create an inverted person-name index.

You need to upload your map/reduce application software to S3, as you saw in the last section. Here, I use the mb option for s3cmd and then the put option to create an S3 bucket that holds both my Ruby scripts and my JAR file containing the Java version of the inverted person-name index creator:

```
s3cmd mb s3://web3-book-map-reduce
s3cmd put namefinder.jar s3://web3-book-map-reduce/
s3cmd put peoplereduce.rb s3://web3-book-map-reduce/
s3cmd put peoplemap.rb s3://web3-book-map-reduce/
```

You can see that the size of the uploaded Ruby map script is huge because it has 100,000 lines of code (mostly data literals):

```
$ s3cmd ls s3://web3-book-map-reduce
2009-04-13 22:07     544248   s3://web3-book-map-reduce/namefinder.jar
2009-04-13 22:08    2852824   s3://web3-book-map-reduce/peoplemap.rb
2009-04-13 22:08        593   s3://web3-book-map-reduce/peoplereduce.rb
```

If you have not already done so, create an Amazon Elastic MapReduce account and log in to the AWS Management Console (https://console.aws.amazon.com/elasticmapreduce/home). The first step is creating a new job flow. I am going to run my Ruby map/reduce scripts that I copied to S3. I enter a name for my job flow and set the job-flow type to Streaming. Entering this information brings up the web page shown in Figure 14-3.

You should notice a few things in Figure 14-3. I added a forward slash to the S3 path name for my input directory web3-book-wikipedia/test/, and I did not add an s3: or an s3n: prefix on this path—that is added automatically. The output path should point to a directory in one of your buckets that does not yet exist, and this path does not need to end with a forward-slash character. I gave the output directory the same name that I assigned to this job on the first web page (which I do not show here). My naming convention, although not required, makes it easier to keep track of jobs and where the output is written. The locations of the map and reduce Ruby scripts are also specified without s3: or s3n: prefixes. There are no required extra arguments. Later, when we monitor a map/reduce run, the AWS Management Console will show you the complete command-line arguments that are used to run Hadoop for your job (see Figure 14-6).

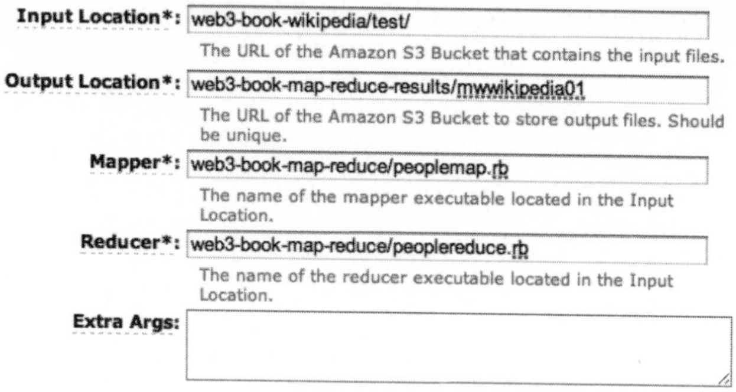

Input Location*: `web3-book-wikipedia/test/`
The URL of the Amazon S3 Bucket that contains the input files.

Output Location*: `web3-book-map-reduce-results/mwwikipedia01`
The URL of the Amazon S3 Bucket to store output files. Should be unique.

Mapper*: `web3-book-map-reduce/peoplemap.rb`
The name of the mapper executable located in the Input Location.

Reducer*: `web3-book-map-reduce/peoplereduce.rb`
The name of the reducer executable located in the Input Location.

Extra Args:

Figure 14-3. After selecting the Streaming option and naming the current job to "mwwiki-pedia01," I entered the Amazon Elastic MapReduce options for input and output data and the location in S3 for the Ruby map and reduce scripts.

Figure 14-4 shows the next screen. The default number of instances is set to four, but for all but very large jobs you can set this to only one instance. You pay a minimum of one hour for an instance ($0.10 per hour is the current cost), so it makes sense to try to use up most of an hour time period in a single run.

Tip If you need to make many small- or medium-size map/reduce runs, you can use Amazon's command-line tools to run a persistent Elastic MapReduce job. Then you can submit multiple runs. See `http://docs.amazonwebservices.com/ElasticMapReduce/latest/DeveloperGuide/index.html?CHAP_Client.html` for directions on using the command-line interface. If this link does not work, search for "Elastic MapReduce ruby command line." Because the hourly charge for an instance is currently just $0.10, I prefer to just use the web interface and pay for a full hour even if a run only takes a few minutes.

The most important option that you need to set is the S3 log path (see Figure 14-4). If you do not set this, you cannot see server and application logs and you'll find it almost impossible to track down errors. I want you to notice that I specified the `s3n:` prefix; this is required, at least in the current version of the AWS Management Console. I also specified just a bucket name. Elastic MapReduce will create a directory for you with the same name as your current job name. Because I also use the job name as my output-directory name, it is easy to find both log files and output for a specific job.

By now you might be thinking that making Hadoop runs using Elastic MapReduce is a lot of work, but you will find that after you run a few successful jobs, you can set up a job very quickly.

After you enter the information on the screen shown in Figure 14-4, you'll see another setup screen that contains a summary of all setup options and a button for submitting the job (see Figure 14-5). After you submit your job, you are offered a link to the job-monitoring page that shows information about recent jobs (see Figure 14-6).

Enter the number and type of EC2 instances you'd like to run your job flow on.

Number of Instances*: `1`

The number of EC2 instances to run in your Hadoop cluster.
If you wish to run more than 20 instances, please complete the limit request form.

Type of Instance*: `Small (m1.small)` ⟐

The type of EC2 instances to run in your Hadoop cluster (learn more about instance types).

⁂ Hide Advanced Options

Amazon S3 Log Path: `s3n://web3-book-map-reduce-logs`

The log path is a location in Amazon S3 where Elastic MapReduce will upload the log files for each step in the job flow. It will take a few minutes after the step has completed for the logs to appear. If you do not specify a path, the log files will not be uploaded.

Amazon EC2 Key Pair: `~ Select EC2 Key Pair ~` ⟐

The Key Pair is the name of an Amazon EC2 Private Key that you have previously created when using Amazon EC2. It is a handle you can use to SSH into the master node of the Amazon EC2 cluster (without a password).

Figure 14-4. *When setting up an Elastic MapReduce job, make sure that you enter an S3 bucket to hold runtime logs.*

Job Flow Name:	mwwikipedia01
Type:	Streaming
Input Location:	s3n://web3-book-wikipedia/test/
Output Location:	s3n://web3-book-map-reduce-results/mwwikipedia01
Mapper:	s3n://web3-book-map-reduce/peoplemap.rb
Reducer:	s3n://web3-book-map-reduce/peoplereduce.rb
Extra Args:	
Number of Instances:	1
Type of Instance:	m1.small
Amazon S3 Log Path:	s3n://web3-book-map-reduce-logs

Figure 14-5. *Elastic MapReduce admin summary screen*

When running map/reduce jobs, I find it convenient to take screenshots of the job-summary web page so I can refer back to them later. The current version of the AWS Management Console does not fill in job-setup information from previous runs for you, so having the summary screens from recent jobs is useful. I also find it useful to keep an open text file that serves as a work log for runs: I can paste the text from the summary web page to my work log as a more permanent record.

Figure 14-6 shows the job-monitoring web page immediately after completion of my "mwwikipedia01" job. You can see that this job took nine minutes to run on one small server instance. Clicking any job in the upper panel shows job detail in the lower panel.

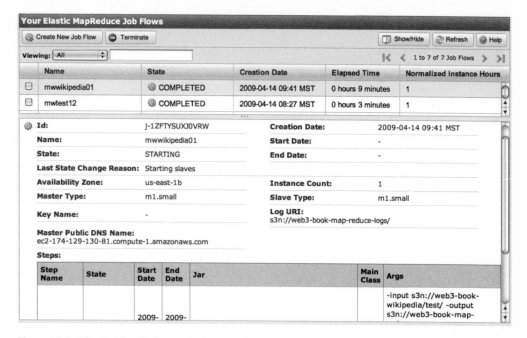

Figure 14-6. *Elastic MapReduce admin console for monitoring map/reduce jobs*

My directions in this section will save you time setting up and running your first few map/reduce data-processing jobs, so you might not need the Amazon Elastic MapReduce Getting Started Guide (http://docs.amazonwebservices.com/ElasticMapReduce/2009-03-31/ GettingStartedGuide/). However, you will want to eventually read through the developer and API guides (http://docs.amazonwebservices.com/ElasticMapReduce/latest/DeveloperGuide/ index.html).

Wrapup

Map/reduce algorithms provide a relatively easy way to scale some types of data-processing tasks to run on a cluster of servers. Map/reduce has been useful for data-processing jobs like machine learning and creating inverted indices for search engines. I recommend that you set up a Hadoop installation on your laptop and try writing map/reduce-based applications for your specific data-processing requirements. Performing development with small input data sets should suffice until you have demonstrated that your map/reduce functions solve your processing problem or that you need to look at other approaches.

Using map/reduce can help you process very large data sets inexpensively. Good strategies include using available servers at nonpeak hours when they are lightly loaded, or purchasing cloud services like Elastic MapReduce.

I will use the Wikipedia inverted person-name index in the next chapter, in which I'll develop an example information portal in Rails.

CHAPTER 15

∎∎∎

Building Information Web Portals

In this chapter, I'll use Rails to develop two information web portals. The first is simple: it uses the inverted person-name index I created in the last chapter using Hadoop to provide local search for people's names on Wikipedia. The second example implements an information-processing portal that is intended for personal use or for use in a small workgroup. This second example allows you to register "interesting things" that can be local document files and web URLs. When you browse your interesting things, you will see similar information sources. You can export both data and associations from the web portal to JSON format using a web-service API, and you can perform SPARQL queries on that data.

The first example web portal is admittedly not a real-world example; rather, its purpose is to give you ideas for your own projects. Creating your own search indices as I did in Chapter 14 and providing local search of your organization's information sources provides highly customized results, unlike Web-wide search engines.

∎**Note** Using Nutch (see Chapter 9) is a good alternative to creating your own search indices with a custom web interface, as I am implementing here.

My hope is that the second web portal—which I'll call the Interesting Things application—will be something you want to experiment with and perhaps customize for your own needs. I have written similar web applications for managing interesting things in Common Lisp and Java. This is an ongoing research interest for me, and I think that you will have fun with the second example.

Searching for People's Names on Wikipedia

The first example web application is simple, but it demonstrates some techniques you might want to implement in your larger Rails web projects. I am using the inverted person-name index that I created in Chapter 14 to let users locally search for names appearing on Wikipedia. (See the section "Running the Ruby Map/Reduce Example Using Amazon Elastic MapReduce" in that chapter.) The output from the map/reduce run resides in the file src/part4/part-00000 (you can find the code samples for this chapter in the Source Code/Download area of the Apress web site). By controlling the information being indexed, you can offer users much more targeted and relevant search results. While I am using search for people's names as an example, your applications might require search for elements such as manufacturing part numbers, descriptions of medical procedures, names of tourist destinations, and so on.

I create a new Rails project and generate a model class that contains attributes for names and web URLs:

```
rails wikipedia_name_finder_web_app
cd wikipedia_name_finder_web_app/
script/generate model NameLink
```

I am going to store data from the inverted person-name index in a MySQL database, so I modified my database.yml file accordingly and set up a database migration to read Hadoop result files generated in the Chapter 14 examples.

■**Tip** You should use compression when storing large amounts of data. I prefer the ZIP file format because it supports multiple internal file entities, unlike gzip, which compresses a single data stream and requires a second tool such as tar to provide internal structure in compressed data archives.

In Chapter 14, I sampled only 6,000 random Wikipedia pages, locating about 22,000 names. This small result set was written to a single Hadoop result file named part-00000. But for real applications, you are likely to process much more data and you will have multiple output files like part-00000, part-00001, and so on. I wrote the database-migration code in this example to process all text files in the ZIP file wikipedia_name_finder_web_app/db/ mapreduce_results.zip and to add the data to a local MySQL database.

I edit the migration file for the NameLink class so it reads all the files from mapreduce_ results.zip into a table called name_links. I add the Ruby code for this calculation in the self.up method immediately after creating the name_links table:

```
require 'zip/zip' # not currently compatible with Ruby 1.9.1

class CreateNameLinks < ActiveRecord::Migration
  def self.up
    create_table :name_links do |t|
      t.column :name, :string
      t.column :url,  :string
    end
    # read Hadoop mapreduce output files in the specified ZIP file:
    ActiveRecord::Base.transaction do
      Zip::ZipInputStream::open("db/mapreduce_results.zip") { |io|
        while (entry = io.get_next_entry)
          io.read.each_line {|line|
            name, urls = line.strip.split("\t")
```

```
          # Hadoop will make sure that all URLs for a given name are on the same
          # line so we can avoid a database lookup to prevent duplicates by simply
          # removing duplicates in the array 'urls':
          urls.split(" ").uniq.each {|url|
            NameLink.create(:name => name, :url => url) if name.length < 25
          }
        }
      end
    }
  end
end

  def self.down
    drop_table :name_links
  end
end
```

I will be making search queries on the column name. You might have noticed that I did not add an index to the name column in the name_links table. This is because I will only be performing matching queries with a wildcard-matching character at the beginning and end of whatever the user types into an AJAX-enabled text search field. Indexing the column name does not improve the performance of queries like this:

```
select * from name_links where name like '%Smith%' limit 2;
```

Indexing the name column does provide better performance if you do not have a wildcard-matching character at the beginning of the search string:

```
select * from name_links where name like 'Betsy%' limit 2;
```

I want to match the characters in the search input field, as the user types them. I want to match anywhere in the name column value, so creating an index incurs unnecessary overhead that provides no performance increase.

■**Note** You might consider using external search tools such as Lucene or Sphinx (see Chapter 9) for much better search performance. However, with these tools you will see results for entire words—users will not see partial-word matches. Sphinx and Lucene can efficiently match partial words starting in the first character position, similar to creating an index in MySQL and only specifying a postfix wildcard character like 'Betsy%'. But that's not what I want for this application.

I use the auto_complete Rails plugin in the next two sections to use the name_links table to dynamically create a completion list while a user is typing characters into a search field.

Using the auto_complete Rails Plugin with a Generated auto_complete_for Method

The functionality for list auto-completion was removed from Rails at version 2.0, so you now need to install it as a separate plugin using this command:

```
script/plugin install auto_complete
```

I am going to use this plugin "out of the box" in this section and then customize its use in the next section. After I generated a controller for search with a method index, I edited the generated class by adding two lines:

```
class SearchController < ApplicationController
  auto_complete_for :name_link, :name, :limit => 25          # added
  protect_from_forgery :only => [:create, :update, :destroy]  # added
  def index
  end
end
```

The plugin added the method auto_complete_for that I use to automatically generate an AJAX handler method called auto_complete_for_name_link_name. The name of this method is derived from the model-class name and the attribute name that I specified in the call to auto_complete_for. This generated method is called asynchronously by the JavaScript that is included on the search web page. The second statement that I added is required to permit AJAX calls to be sent back to the web application through HTTP GET calls. Starting with Rails 2.0, if you forget to add this statement, nothing will happen when the JavaScript on your web page makes AJAX calls back to the server. To see how this plugin works, start by looking at my edited version of the view template app/views/search/index.html.erb:

```
<%= javascript_include_tag :all, :cache => true %>

<h1>Local Search for Names on Wikipedia</h1>
<p>Use auto-complete to find a name and then visit the Wikipedia page:</p>

<%= text_field_with_auto_complete :name_link, :name, :size=>20 %>
```

The first statement ensures that the standard Rails JavaScript libraries are available on the HTML page that this view generates. The last statement uses the class name and attribute name that I used in my call to the method auto_complete_for in the search controller as arguments for calling the method text_field_with_auto_complete supplied by the plugin.

After making these changes, you can run the web application and see the popup completion list as shown in Figure 15-1.

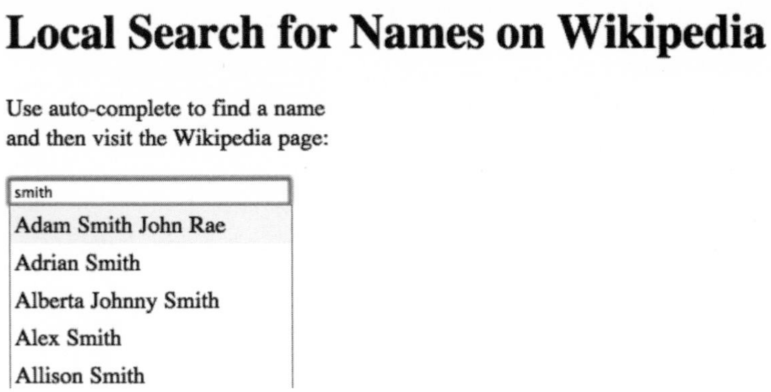

Local Search for Names on Wikipedia

Use auto-complete to find a name
and then visit the Wikipedia page:

Figure 15-1. *Using the default generated auto_complete_for method*

I have not assigned any actions for selecting a name on the completion list. I will do that
in the next section, when I write a custom AJAX handler method in the search controller.

Using the auto_complete Rails Plugin with a Custom auto_complete_for Method

My use of the auto_complete plugin in the preceding section is probably the way most people
use it. However, I find that *not* calling auto_complete_for in the controller class and writing my
own AJAX handler method instead is often useful because it gives me the flexibility of generating arbitrary HTML fragments to return to the JavaScript in a web page.

For this example, I want the completion list to show the names matching the characters a
user is typing into a search field and also the target URL (see Figure 15-2). I edited my search
controller to comment out the call to auto_complete_for and implemented my own version of
the method auto_complete_for_name_link_name:

```
require 'cgi'

class SearchController < ApplicationController
  #auto_complete_for :name_link, :name, :limit => 25
  protect_from_forgery :only => [:create, :update, :destroy]
  def index
  end
```

The auto_complete_for_name_link_name method is called repeatedly as a user types text
into the search field on the web page. I first make sure that there is a value (another hash table)
in the params hash table for the name_link field and that this hash table has a value for the key
name. You might find it useful to print out the value of params['name_link']['name'] and watch
the Rails console output while you enter characters on the search web page. When I wrote
this method, I also printed out the HTML fragment that I was generating to send back to the
JavaScript on the web page. The native SQL query that I execute returns values for a name and
URL for each query result. I build an HTML fragment that contains a list of match results:

```
  def auto_complete_for_name_link_name
    if params['name_link'] && name = params['name_link']['name']
      results = ActiveRecord::Base.connection.execute("select name, url from " +
                              "name_link where name like '%#{name}%' limit 15")
      html = "<ul>"
      results.each {|pname, plink|
          html << ("<li>#{pname}<br/>  <a href=\"#{plink}\"> \
                              #{CGI.unescape(plink)}</a></li>") if plink.length<81}
      html << "</ul>"
      render :inline => html
    end
  end
end
```

Notice that in each list element I construct, I add the person name followed by a link to the URL of the original Wikipedia article. To make the text in the link look better, I CGI-unescape it to convert encoded characters back to a more readable form. This example is shown in Figure 15-2.

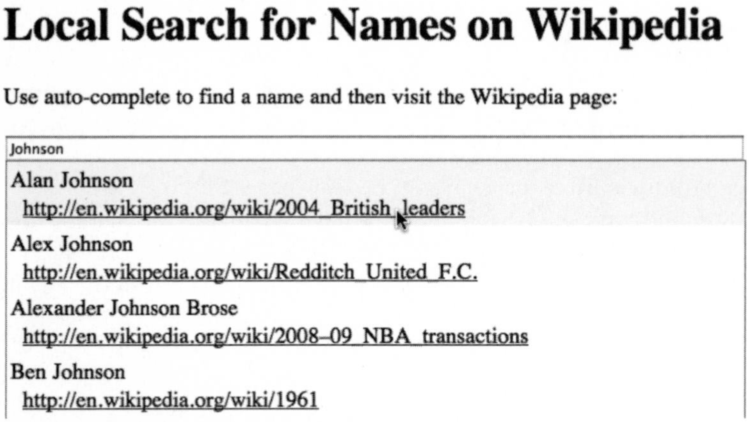

Local Search for Names on Wikipedia

Use auto-complete to find a name and then visit the Wikipedia page:

Johnson
Alan Johnson
http://en.wikipedia.org/wiki/2004_British_leaders
Alex Johnson
http://en.wikipedia.org/wiki/Redditch_United_F.C.
Alexander Johnson Brose
http://en.wikipedia.org/wiki/2008–09_NBA_transactions
Ben Johnson
http://en.wikipedia.org/wiki/1961

Figure 15-2. *Using the custom auto_complete_for method*

You might be wondering about the performance issues associated with making a database query to handle each AJAX call. Because I am performing partial-string matches, MySQL makes a sweep through each row, performing a match on the name column. I benchmarked the database query calls on my MacBook, and they took between 1.5 and 2.5 milliseconds each. If you have a very large number of rows, you can use Sphinx as an alternative (see Chapter 9). For some applications, you can cache data in the browser and implement local search tips in JavaScript.

This was a simple web application, but hopefully you learned some new techniques that you hadn't seen before. The example developed in the remainder of this chapter demonstrates more web-application development techniques that I think you will find useful.

A Personal "Interesting Things" Web Application

The second example application is a simpler version of a long-term research experiment that I implemented in Common Lisp (using the Allegro Webactions framework and the Dojo JavaScript library). Don't worry, though: I implement the example in this section using Rails and the standard Rails JavaScript libraries. I think you will find the source code for this example easy to work with for your own experiments.

This web application lets you register both local files and web URLs. The text from these original information sources is processed to provide automatically generated summaries and category assignments. You can modify the automatically assigned summaries and category assignments, and your changes are permanently stored. A batch process that you can start from an admin web page looks for and records similarities between information sources, or "interesting things." As you will see, the web application supports search and browsing by category.

As you read local documents and web pages, you can record the web page URLs and upload copies of local documents using this example web application. This web app has four main pages:

- *Home*: Lets you add local files and web URLs as interesting things, and lets you perform local search

- *Categories*: Lets you browse interesting things by category

- *Admin*: Lets you start and stop the Sphinx server, reindex, and run a batch job for finding similarities between interesting things

- *Web Service Info*: Provides information for using the web-service APIs

My approach to assigning categories in this example resembles the approach I used in Chapter 3. I use pretrained categories, whose data resides in the directory db/categories_as_text/. Users of this web application can manually change assigned categories for the interesting things they have registered with the system.

■**Note** In addition to being derived from my Common Lisp–based system, this example is based loosely on my new project web site http://www.mythingsofinterest.com/, which runs on the Java edition of Google App Engine.

In the next section, I describe the implementation of the back-end processing code for this example web application. Later, I'll cover the implementation of the web interface. I am going to include some screenshots of the final version of the web application as I discuss back-end processing that provides the data to the UI.

Back-End Processing

For any web application, I like to write the back-end code first. Once the back end is written and tested, the web user interface is much easier to write. I am going to use Sphinx to provide search (see Chapter 9 for details on setting up Sphinx).

I start by generating a new Rails application and the four model classes that we will need:

```
rails interesting_things_web_app
cd interesting_things_web_app/
script/generate model document
script/generate model similar_link
script/generate model document_category
script/generate model category_word
```

The classes SimilarLink, DocumentCategory, and CategoryWord are used as-is, with no changes to the code required. The class Document contains all data attributes for text processing and search. The controller classes contain very little application-specific code; it is mostly implemented in Document.

■**Note** You need to install the stemmer gem: gem install stemmer.

To show you the data models, I list the migration files for all four model classes (with some details left out):

```
create_table :documents do |t|
  t.string :uri, :null => false
  t.string :original_source_uri
  t.string :summary
  t.text :plain_text  # use a Sphinx index on this attribute
end
create_table :similar_links do |t|
  t.integer :doc_id_1, :null => false
  t.integer :doc_id_2, :null => false
  t.float     :strength, :null => false
end
create_table :document_categories do |t|
  t.integer :document_id
  t.string    :category_name
  t.float     :likelihood # of category being correct. range: [0.0, 1.0]
end
```

The migration for the class CategoryWord is a little more complicated because I initialize the database table by reading the text files in the directory db/categories_as_text:

```
create_table :category_words do |t|
  t.string :word_name
  t.string :category_name
  t.float  :importance
end
Dir.entries("db/categories_as_text").each {|fname|
  if fname[-4..-1] == '.txt'
      File.open("db/categories_as_text/#{fname}").each_line {|line|
        word, score = line.split
        CategoryWord.create(:word_name => word,
                                   :category_name => fname[0..-5],
                                   :importance => score.to_f * 0.1)

      }
  end
}
```

I often store word associations for categories in in-memory hash tables, but I want to keep the memory footprint as small as possible for each Mongrel process running this web application. This overhead for hitting a local database occurs only when local documents and web URIs are registered with the system.

Except for some Rails AJAX code you'll see later, most of this app's interesting code is in the Document class, so I'll list the entire document.rb file interspersed with explanations:

```
require 'stemmer'
require 'fileutils'
require 'open-uri'
```

I will need common words that include a trailing period for splitting text into sentences:

```
HUMAN_NAME_PREFIXES_OR_ABREVIATIONS = ['Mr.', 'Mrs.', 'Ms.', 'Dr.', 'Sr.', ➥
'Maj.', 'St.', 'Lt.', 'Sen.', 'Jan.', 'Feb.', 'Mar.', 'Apr.', "Jun.", 'Jul.', ➥
'Aug.', 'Sep', 'Oct.', 'Nov.', 'Dec.']
DIGITS = ['0', '1', '2', '3', '4', '5', '6', '7', '8', '9']
```

The Document class has one-to-many associations with the classes DocumentCategory and SimilarLink. I will need access to noise-word (or stop-word) stems (see Chapter 3 for a discussion about word stemming). I will also need a list of all active category names (variable @@cat_names) in both the batch program script/find_similar_things.rb and the Rails UI implementation:

```
class Document < ActiveRecord::Base
  has_many :document_categories, :dependent => :destroy
  has_many :similar_links, :dependent => :destroy
  attr_accessor :calculated_summary
  attr_accessor :calculated_categories
  @@cat_names = nil
  @@noise_stems = nil
```

The method define_index was added to the ActiveRecord base class when I installed the Thinking Sphinx plugin (see Chapter 9 for details). Here I declare that Sphinx should index the column plain_text in the database table documents:

```
define_index do
  indexes [:plain_text]
end
```

The following class method, Document.from_local_file, is a factory for creating a new class instance from a local file. I use the command-line utilities pdftotext and antiword to extract text from PDF and Microsoft Word files. (Read more about these utilities in Chapter 1; you can find them on the Web if you do not already have them installed. There are installers for Linux, Mac OS X, and Windows.) I call the methods semantic_processing and calculated_summary to calculate attribute values and save the new document instance to the database:

```
def Document.from_local_file file_path, original_source_uri
  index = original_source_uri.rindex(".")
  file_extension = original_source_uri[index..-1]
  permanent_file_path = "db/document_repository/d#{Date.new.to_s}-" +
                                    "#{rand(100000).to_s}#{file_extension}"
  plain_text = ''
  if file_extension == ".txt"
    FileUtils.cp(file_path, permanent_file_path)
    plain_text = File.new(permanent_file_path).read
  elsif file_extension == ".pdf"
    `pdftotext #{file_path} #{permanent_file_path}`
    plain_text = File.open(permanent_file_path, 'r').read
  elsif file_extension == ".doc"
    plain_text = `antiword #{file_path}`
    File.open(permanent_file_path, 'w') {|out| out.puts(plain_text)}
  end
  doc = Document.new(:uri => permanent_file_path,
                          :plain_text => plain_text,
                          :original_source_uri => original_source_uri)
  doc.semantic_processing
  doc.summary = doc.calculated_summary
  doc.save! # need to set doc's id before the next code block:
  score = 0.5
  doc.calculated_categories.each {|category|
    doc.category_assigned_by_nlp(category, score) if score > 0.1
    score *= 0.5 # scores are ordered, so decrease the value for next score
  }
end
```

The Document.from_web_url method is similar to the Document.from_local_file method, but instead of reading a local file I fetch the contents of a web page:

```
def Document.from_web_url a_url
  begin
    plain_text = open(a_url).read.gsub(/<\/?[^>]*>/, " ").gsub(➥
```

```
' ', ' ').gsub(' ', ' ')
      return false if plain_text.index("File Not Found")
      return false if plain_text.index("404 Not Found")
      return false if plain_text.index("Error: getaddrinfo")
      file_extension = '.html'
      permanent_file_path = "db/document_repository/d#{Date.new.to_s}-" +
                                      "#{rand(100000).to_s}#{file_extension}"
      doc = Document.new(:uri => permanent_file_path,
                                  :plain_text => plain_text,
                                  :original_source_uri => a_url)
      doc.semantic_processing
      doc.summary = doc.calculated_summary
      doc.save! # need to set doc's id before the next code block:
      score = 0.5
      doc.calculated_categories.each do|category|
        doc.category_assigned_by_nlp(category, score) if score > 0.1
        score *= 0.5
      end
    rescue
      puts "\n** Document.from_web_url:  #{a_url}  Error: #{$!}"
      return false
    end
    true # OK
  end
```

The following class method, Document.get_all_category_names, returns a list of category names. This list is cached, so subsequent calls to this method do not need to access the database:

```
  def Document.get_all_category_names
    return @@cat_names if @@cat_names
    result = ActiveRecord::Base.connection.execute(
       "select distinct category_name from category_words order by ➥
category_name asc")
    @@cat_names = []
    result.each {|x| @@cat_names << x[0]} # MySQL::Result class has no
                                   # collect method
    result.free
    @@cat_names
  end
```

The Document.get_noise_word_stems class method returns a list of noise words (stop words) that are stemmed. This list is also cached, so subsequent calls to this method do not need to reread the text file db/stop_words.txt:

```
  def Document.get_noise_word_stems
    return @@noise_stems if @@noise_stems
    @@noise_stems = []
    f = File.open('db/stop_words.txt')
```

```
      f.read.split("\n").each {|line|
        @@noise_stems << line.strip.stem
      }
      f.close
      @@noise_stems
    end
```

The method `semantic_processing` processes plain text to determine possible categories for the text contents and creates a summary. This code resembles what I wrote in Chapter 3:

```
def semantic_processing
  category_names = Document.get_all_category_names
  breaks = get_sentence_boundaries(plain_text)
  word_stems = plain_text.downcase.scan(/[a-z]+/).collect {|word| word.stem}
  scores = Array.new(category_names.length)
  category_names.length.times {|i| scores[i] = 0}
  word_stems.each {|stem|

    CategoryWord.find(:all, :conditions => {:word_name => stem}).each {|cw|
      index = category_names.index(cw.category_name)
      scores[index] += cw.importance
    }
  }
  slist = []
  category_names.length.times {|i|
                   slist << [scores[i], category_names[i]] if scores[i] > 0}
  slist = slist.sort.reverse
  @calculated_categories =
      slist[0..category_names.length/3+1].collect {|score, cat_name| cat_name}

  best_category = @calculated_categories[0]
  sentence_scores = Array.new(breaks.length)
  breaks.length.times {|i| sentence_scores[i] = 0}
  breaks.each_with_index {|sentence_break, i|
    tokens = plain_text[sentence_break[0]..sentence_break[1]]. ➥
                               downcase.scan(/[a-z]+/).collect {|tt| tt.stem}
    tokens.each {|token|
      CategoryWord.find(:all,
                  :conditions =>
                      {:word_name => token,
                       :category_name => best_category}).each {|cw|
                                    sentence_scores[i] += cw.importance }
    }
    sentence_scores[i] *= 100.0 / (1 + tokens.length)
  }

  score_cutoff = 0.8 * sentence_scores.max
  summary = ''
```

```
sentence_scores.length.times {|i|
  if sentence_scores[i] >= score_cutoff
    summary << plain_text[breaks[i][0]..breaks[i][1]] << ' '
  end
}
@calculated_summary = summary.strip
end
```

The following two methods update the database with categories set by the NLP code or input from a user. Both use the private helper method category_assigned_helper that you will see later:

```
def category_assigned_by_nlp category, likelihood
  category_assigned_helper(category, false, likelihood)
end
def category_assigned_by_user category
  category_assigned_helper(category, true, 1.0)
end
```

The method similarity_to calculates a similarity value in the range [0, 1.0] between this document and another document. A list of unique word stems is created for both documents, and the size of the intersection of these lists is scaled by the number of words in each document. I calculate similarity by comparing the number of common word stems in two documents with the number of words in both documents. I scale this calculation by squaring the word-intersection count and dividing it by the number of words in the first document times the number of words in the second document:

```
def similarity_to another_document
  noise = Document.get_noise_word_stems
  text_1 = ((plain_text.downcase.scan(/[a-z]+/).collect {|word|
                                              word.stem}) - noise).uniq
  text_2 = ((another_document.plain_text.downcase.scan(/[a-z]+/).collect {|word|
                                              word.stem}) - noise).uniq
  f1 = (text_1 & text_2).length.to_f
  f2 = text_1.length.to_f
  f3 = text_2.length.to_f
  (f1 * f1) / (f2 * f3)
end
```

The method get_similar_document_ids returns the IDs of similar documents. Each ID is paired with the similarity value. This is a simple database lookup because these values are calculated in the batch program scripts/find_similar_things.rb:

```
def get_similar_document_ids
  SimilarLink.find(:all, :order => :strength,
                        :conditions => {:doc_id_1 => id}).collect {|x|
                                              [x.doc_id_2, x.strength]}
end
```

The following private methods write document-category associations to the database and calculate the location of sentence boundaries (see Chapter 3 for a discussion of this operation):

```
private  # PRIVATE:
def category_assigned_helper category, by_user, likelihood
  results = DocumentCategory.find(:first,
                :conditions => {:document_id => id, :category_name => category})
  if !results
    DocumentCategory.create(:document_id => id,
                                     :category_name => category,
                                     :set_by_user => by_user,
                                     :likelihood => likelihood)
  end
end

def get_sentence_boundaries text
  boundary_list = []
  start = index = 0
  current_token = ''
  text.each_char {|ch|
    if ch == ' '
      current_token = ''
    elsif ch == '.'
      current_token += ch
      if !HUMAN_NAME_PREFIXES_OR_ABREVIATIONS.member?(current_token) &&
         !DIGITS.member?(current_token[-2..-2])
        boundary_list << [start, index]
        current_token = ''
        start = index + 2
      else
        current_token += ch
      end
    elsif ['!', '?'].member?(ch)
        boundary_list << [start, index]
        current_token = ''
        start = index + 2
    else
      current_token += ch
    end
    index += 1
  }
  boundary_list
end
end
```

The document similarity is calculated in the batch program scripts/find_similar_things.rb that I'll list here, interspersed with explanations. I'll run the web application in development mode, which is better to experiment with (but you can change the MODE constant if you want to run in production mode):

```ruby
#!/usr/bin/env ruby
MODE = 'development'
```

The class Document requires the Thinking Sphinx plugin to load without error:

```ruby
$: << File.dirname(__FILE__) + '/../vendor/plugins/thinking-sphinx/lib'
$: << File.dirname(__FILE__) + '/../app/models/'
require 'thinking_sphinx' # only required so model files will load
require 'activerecord'
require 'document'
require 'similar_link'
```

I read the database.yml file to get the name of the database:

```ruby
database_config = YAML.load_file(File.dirname(__FILE__) + "/../config/database.yml")
database_name = database_config[MODE]['database']
database_adapter = database_config[MODE]['adapter']


puts "\n****** Starting script find_similar_things.rb *******\n"
puts "** Database adapter: #{database_adapter}"
puts "** Database name:    #{database_name}"
```

Next, I need to open a database connection and fetch the IDs of all documents in the database:

```ruby
ActiveRecord::Base.establish_connection(:adapter  => database_adapter,
                                        :database => database_name)
doc_ids = Document.find(:all, :select=>'id').collect {|doc| doc.id}
```

I have a double nested loop of document IDs, so this script runs at O(n^2) where n is the number of documents:

```ruby
doc_ids.each {|id_1|
  doc_1 = Document.find(id_1)
  ss = doc_1.get_similar_document_ids
  doc_ids.each {|id_2|
    if id_1 != id_2
      links = SimilarLink.find(:first, :conditions => {:doc_id_1 => id_1,
                                                       :doc_id_2 => id_2})

      if !links
        doc_2 = Document.find(id_2)
        similarity = doc_1.similarity_to(doc_2)
        puts "similarity: #{similarity} #{doc_1.original_source_uri} " +
             " #{doc_2.original_source_uri}"
        if similarity > 0.1
          SimilarLink.new(:doc_id_1 => id_1, :doc_id_2 => id_2,
                          :strength => similarity).save!
        end
```

```
      end
    end
  }
}
```

If you have a very large number of documents, consider using the Ruby gem `clusterer` that implements K-means clustering (see Chapter 3).

You can run this `find_similar_things.rb` script periodically from a cron job. While experimenting with the web app, you can also run it from the Admin web page using the "Find similar stuff" link (see Figure 15-3).

Home I Categories I Admin I Web Service Info

Run Sphinx indexer

Start Sphinx (disabled: Sphinx already running)

Stop Sphinx

Find similar stuff

```
******* Starting script find_similar_things.rb *******
** Database adapter: mysql
** Database name:     interesting_things_development
similarity: 0.0 test.txt http://knowledgebooks.com/
similarity: 0.0 test.txt http://markwatson.com
similarity: 0.0 http://knowledgebooks.com/ test.txt
similarity: 0.126807219590725 http://knowledgebooks.com/ http://markwatson.com
similarity: 0.0 http://markwatson.com test.txt
similarity: 0.126807219590725 http://markwatson.com http://knowledgebooks.com/
```

Figure 15-3. *Running the find_similar_things.rb script from the Admin web page*

You periodically want to run the Sphinx indexer that rotates indices; you can do so while performing search with Sphinx. Cron is a good tool for running periodic processes like the indexer. You can manually run the indexer and other Sphinx operations using the rake tasks:

```
rake thinking_sphinx:index
rake thinking_sphinx:restart
rake thinking_sphinx:running_start
rake thinking_sphinx:start
rake thinking_sphinx:stop
```

The Admin web page shown in Figure 15-3 also allows you to start and stop Sphinx and run the indexer. If you start Sphinx using the Admin page and stop running the Mongrel process that spawned the indexing process, then the Sphinx child process is terminated.

■**Warning** Sphinx with the Thinking Sphinx plugin does not automatically index new rows added to a database. If you need instant indexing of new objects, consider using Solr and the `acts_as_solr` plugin (see Chapter 9).

Except for some use of AJAX, the Rails web interface is fairly simple. I cover the implementation of the Rails UI in the next section.

Rails User Interface

As I mentioned earlier, the web user interface has four main pages: Home, Categories, Admin, and Web Service Info. So I start by creating a controller for each page and an index method and index view for each controller:

```
script/generate controller home index
script/generate controller categories index
script/generate controller admin index
script/generate controller info index
```

I will start with the Home page, shown in finished form in Figure 15-4. The Home page allows users to add documents from local files and web URLs, search for documents, view a selected document's original text, view and edit its summary and assigned categories, and view a list of similar things associated with it.

Home | Categories | Admin | Web Service Info

| java jsp | (Search) |

Results:

http://knowledgebooks.com/ ◉visit site
http://markwatson.com ◉visit site

Selected interesting thing: http://knowledgebooks.com/

"AI text mining tools for Java, Ruby, and Common Lisp"

◉View text ◉Edit summary ◉Edit categories ◉View similar things

Notice something new:

Load a local file:

(Choose File) no file selected
(Submit)

Load a web page by URL:

URL: [   ]
(Submit)

Figure 15-4. *Home page after searching for "java jsp" and selecting the first search result*

■Note I used the free circular icon collection from `http://prothemedesign.com/`
`free-webdesign-tools/circular-icons/` for this example web application.

I use one view template for the Home page and a partial `_results.html.erb` that is also used by the categories controller. Handling file uploads is problematic in Rails applications if you want to stay on the original page. There is a well-known and often used trick for using a hidden iFrame as a form target (for references, search the web for "Rails file upload iFrame"), but I have had only partial success with this trick for some web browsers. Instead, I prefer to perform actual file uploads with nginx and use JavaScript to call out to nginx for download progress. nginx, a high-performance HTTP server and reverse proxy, is my preferred server for deploying Rails applications. I don't cover Rails deployment in this book, so I refer you to the article at `http://brainspl.at/articles/tag/nginx` that provides good instructions for configuring nginx and using the nginx upload module (for details on the nginx upload module, see `http://www.grid.net.ru/nginx/upload.en.html`).

Because I want this example to run using just Mongrel for casual deployments (deployments that don't require nginx), I implemented local file uploads and remote web-page fetches with actions implemented by the home controller methods `upload` and `get_url`:

```
<form action="/home/upload" method="post" enctype="multipart/form-data">
<form action="/home/get_url" method="post">
```

After you upload local files and fetch web pages, the application shows status web pages for two seconds before using an HTTP redirect back to the Home page. Not a perfect solution, but it uses standard HTML and HTTP, so it should work for all browsers.

Viewing a Selected Document's Text on the Home Page

In the bottom-left part of Figure 15-4, you see four links that show details of documents selected from search results: "View text," "Edit summary," "Edit categories," and "View similar things." Figures 15-5, 15-6, 15-7, and 15-8 show the screens that you see when you click each of these options. The partial `app/views/home/_results.html.erb` generates the HTML for these four links and their resulting pages.

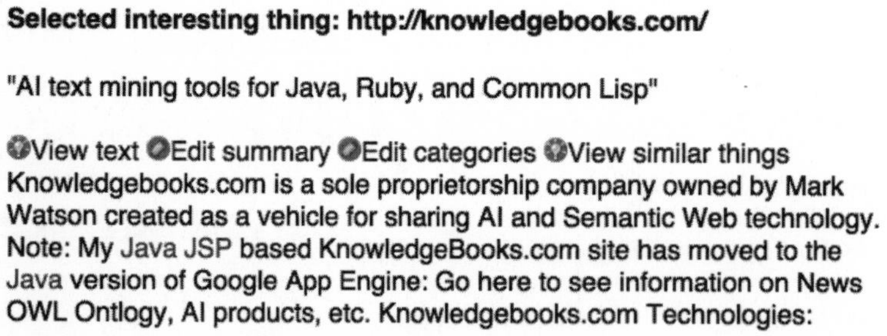

Selected interesting thing: http://knowledgebooks.com/

"AI text mining tools for Java, Ruby, and Common Lisp"

⊕View text ⊖Edit summary ⊖Edit categories ⊕View similar things
Knowledgebooks.com is a sole proprietorship company owned by Mark
Watson created as a vehicle for sharing AI and Semantic Web technology.
Note: My Java JSP based KnowledgeBooks.com site has moved to the
Java version of Google App Engine: Go here to see information on News
OWL Ontlogy, AI products, etc. Knowledgebooks.com Technologies:

Figure 15-5. *This is how the Home page appears after a user clicks "View text" for a selected document. Search terms appear in a red font.*

The following listing shows the part of the fragment `app/views/home/_results.html.erb` that generates the "View text," "Edit summary," "Edit categories," and "View similar things" links (see Figures 15-5 through 15-8). The following snippet also shows the template for generating the view-text display with search words highlighted in a red font. I use the JavaScript helper function `link_to_function` to generate links that toggle the visibility of HTML `div` elements that are hidden. Displaying text from a document is easy. I fetch the attribute `plain_text` from a document object and use the string `gsub` method with a regular expression and code block to highlight search words:

```
<h4>Selected interesting thing: <%=@doc.original_source_uri%></h4>
&quot;<%=@doc.summary[0..80]%>&quot;<br/><br/>
```

```
<image src="images/bulb.png"/><%= link_to_function "View text", ➥
"$('view_text').toggle();"%>
<image src="images/edit.png"/><%= link_to_function "Edit summary", ➥
"$('generated_summary').toggle();"%>
<image src="images/edit.png"/><%= link_to_function "Edit categories", ➥
"$('category_edit').toggle();"%>
<image src="images/bulb.png"/><%= link_to_function "View similar things", ➥
"$('similar_things').toggle();"%>

<div id="view_text"style="display:none; width:65%">
<%=
  hilite_words = session['search_text'].split
  text = @doc.plain_text
  hilite_words.each {|word|
     text = text.gsub(/#{word}\b/i) {|w| "<font color=\"red\">#{w}</font>"}}
  text
%>
</div>
```

Viewing and Editing the Summary

Figure 15-6 shows the second detail option: letting the user see the summary for a document and make manual edits to the generated summary.

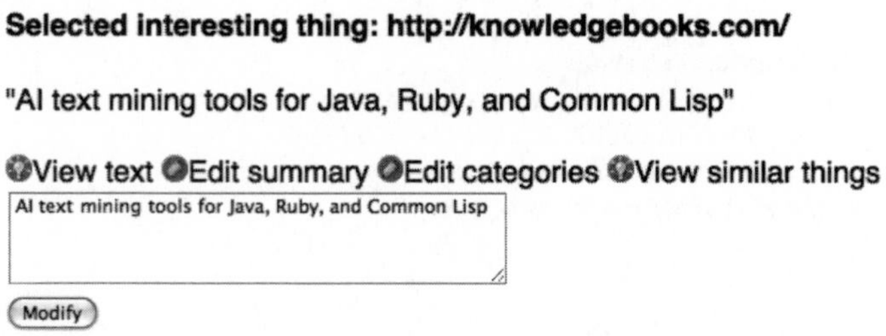

Selected interesting thing: http://knowledgebooks.com/

"AI text mining tools for Java, Ruby, and Common Lisp"

💡View text ✏Edit summary ✏Edit categories 💡View similar things

```
AI text mining tools for Java, Ruby, and Common Lisp
```

Modify

Figure 15-6. *Viewing the summary for a selected document*

The code in the _results.html.erb fragment that generates the detail display in Figure 15-6 is shown in next two code snippets:

```
<div id="generated_summary"style="display:none">
<% remote_form_for :document, @doc, :update => 'status_results', :url => { ➥
:action => "update_summary" } do |f| %>
  <%=text_area(:doc, :summary, :size => "40x4")%>
  <%=hidden_field(:doc, :id)%>
  <br/><%= submit_tag 'Modify' %>
<% end %>
</div>
```

In this display of the summary, I used an editable text area to display the summary and the AJAX helper method `remote_form_for` that calls the home controller method `update_summary`. This controller method updates the database and returns a status message. The JavaScript generated by the `remote_form_for` method uses this status message to update the HTML element with an ID equal to `status_results`. Now is a good time to look at the home controller method `update_summary`:

```
def update_summary
  begin
    doc = params['doc']
    Document.update(doc['id'], :summary => doc['summary'])
    render :text => "Summary update OK"
  rescue
    render :text => "Error: #{$!}"
  end
end
```

Here, the document selected on the Home page is passed in the `params` hash table. I update the database and render message text on the Home page element `<p id="status_results"></p>`.

Viewing and Editing the Assigned Categories

Figure 15-7 shows the category detail for the selected document. Figure 15-7 shows only eight of a total of 27 categories; I cropped the screenshot to save space.

Selected interesting thing: http://knowledgebooks.com/

"AI text mining tools for Java, Ruby, and Common Lisp"

View text Edit summary Edit categories View similar things

computers:	☐ computers_ai:	☑
computers_ai_learning:	☐ computers_ai_nlp:	☑
computers_ai_textmining:	☑ computers_microsoft:	☐
computers_programming_c++:	☐ computers_programming_java:	☑

Figure 15-7. *Home page showing the assigned categories for a selected document*

The code in the `_results.html.erb` fragment that generates the detail display in Figure 15-7 is shown in the following listing, interspersed with explanations. I need to get all category names used in the system, access the database to see which categories are assigned to this document, and display the labeled check boxes with or without a check, as appropriate. If the user clicks the Modify button, then the home controller method `update_categories` is called to update the database. While in general you want to avoid embedded code and loops in Erb templates, displaying the categories does require some extra logic because we want to display three columns of categories. The home controller method `result_detail` defines `@doc` for displaying the selected document text, summary, category list, and view of similar documents. This method also defines a list of all possible document names (`@dcs`) and all possible category names (`@all_cats`):

```
def result_detail
  @doc = Document.find(params[:doc_id])
  @dcs = @doc.document_categories.collect {|dc| dc.category_name}
  @all_cats = Document.get_all_category_names
  render :partial => 'results'
end
```

The following snippet from the partial _results.html.erb displays the categories defined for a document and lets the user edit the assigned categories:

```
<div id="category_edit" style="display:none">
<% remote_form_for :document, @doc, :update => 'status_results',
                                :url => { :action => "update_categories" } do |f| %>
  <%=hidden_field(:doc, :id)%>
  <table>
  <% @all_cats.each_with_index { |cn, count|
      if count == 0  %>
        <tr>
<% end %>
      if @dcs.include?(cn)  %>
        <td><%=cn%>:</td>
        <td><%=check_box_tag(cn, "1", true)%></td>
        <td style="width:5">
<%  else%>
        <td><%=cn%>:</td>
        <td><%=check_box_tag(cn, "0", false)%></td>
        <td style="width:5">
<% end %>
      if (count + 1) % 3 == 0 %>
        </tr>
<% end %>
    } %>
  </table>
  <%= submit_tag 'Modify' %>
<% end %>
</div>
```

The code in this fragment template organized the labeled check boxes into rows of three check boxes each. This view code fragment calls the update_categories controller method that is shown in the next listing:

```
def update_categories
  begin
```

The params hash table contains the document data encoded as a hash. Here, I save this hash and then cycle through the params hash, saving everything that is a category name in the cat_names array:

```
    doc = params['doc']
    cat_names = params.collect {|p|
        p[0] if !['authenticity_token', '_', 'commit', 'controller', 'action', ➥
'doc'].include?(p[0])
    }
    cat_names.delete(nil)
```

I now fetch the current document from the database, clear out all categories currently associated with this document, and define a new set of categories using the names in the array cat_names:

```
    doc = Document.find(doc['id'])
    doc.document_categories.clear
    cat_names.each {|cn| doc.category_assigned_by_user(cn)}
    doc.save!
    render :text => "Categories update OK"
  rescue
    render :text => "Error: #{$!}"
  end
 end
```

Viewing a List of Similar Things

Figure 15-8 shows the detail display for viewing similar documents (or "similar things," in keeping with the theme of the example application).

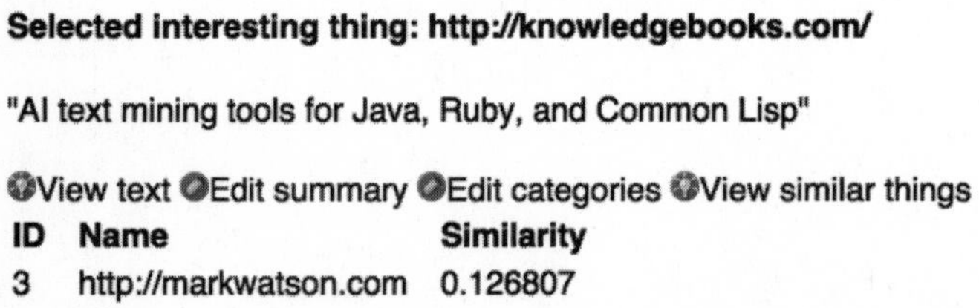

Selected interesting thing: http://knowledgebooks.com/

"AI text mining tools for Java, Ruby, and Common Lisp"

◉View text ◉Edit summary ◉Edit categories ◉View similar things

ID	Name	Similarity
3	http://markwatson.com	0.126807

Figure 15-8. *Home page showing documents similar to the selected document*

The code in the following listing shows the part of the _results.html.erb fragment that generates the detail display in Figure 15-8:

```
<div id="similar_things"style="display:none">
 <table>
    <tr><td><strong>ID</strong></td><td>  
        </td><td><strong>Name</strong></td><td>  
        </td><td><strong>Similarity</strong></td></tr>
<% @doc.get_similar_document_ids.each {|other_doc_id, strength|
    other_doc = Document.find(other_doc_id)
%>
    <tr>
```

```
      <td><%=other_doc.id%></td>
      <td>  </td>
      <td><%=other_doc.original_source_uri.strip%></td>
      <td>  </td>
      <td><%=strength%></td>
    </tr>
<% } %>
  </table>
</div>
```

Implementing the Categories Page

That covers the implementation of our application's Home page. Now I'll implement the Categories page as shown in Figures 15-9 and 15-10.

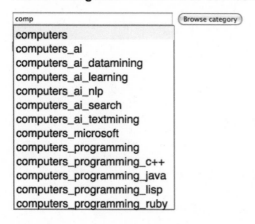

Figure 15-9. *Categories page showing auto-complete functionality for specifying a category name*

I want users to start typing a category name into an auto-complete field (similar to what I implemented in this chapter's first example) and then select one of the displayed categories (see Figure 15-9). At this point, they see a list of interesting things that are assigned that category, sorted according to the likelihood that a category assignment is correct (see Figure 15-10).

As you did in this chapter's first example, use the following command to install the auto-complete functionality as a separate plugin:

```
script/plugin install auto_complete
```

I implemented a custom auto-complete event handler in the category controller that handles AJAX calls from the view:

```
def auto_complete_for_document_category_category_name
  c_names = Document.get_all_category_names
  html = '<ul>'
  if  params['document_category'] &&
      cname = params['document_category']['category_name']
    c_names.each {|cn| html << "<li>#{cn}</li>" if cn.index(cname)}
  end
  html << '</ul>'
  render :inline => html
end
```

Here I am matching the characters that a user has typed against the category names used in the system. Matching category names are added to an HTML list element. I use the same results display for selected documents on the Categories page that I used on the Home page (see Figure 15-10).

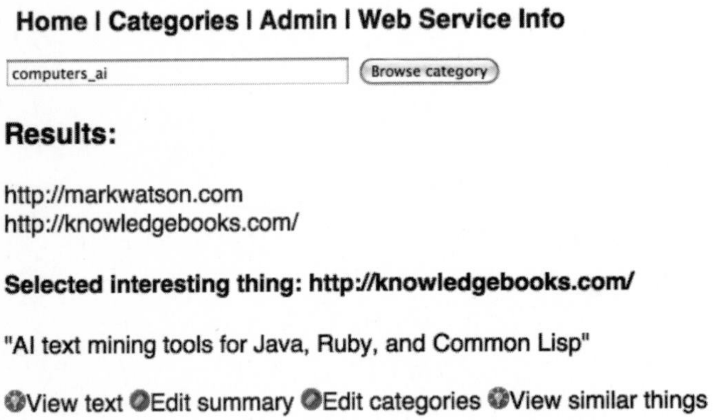

Home | Categories | Admin | Web Service Info

computers_ai   (Browse category)

Results:

http://markwatson.com
http://knowledgebooks.com/

Selected interesting thing: http://knowledgebooks.com/

"AI text mining tools for Java, Ruby, and Common Lisp"

◉View text ◐Edit summary ◐Edit categories ◉View similar things

Figure 15-10. *Categories page after selecting the category "computers_ai" and selecting one of the result links*

The category controller method index checks whether the user has specified a category name. If so, then a list of all documents with this category assignment is added to the array stored in @doc_list that the view template will use:

```
def index
  @doc_list = []
  if  params['document_category'] &&
      cname = params['document_category']['category_name']
    DocumentCategory.find(:all, :order => :likelihood,
                          :conditions => {:category_name => cname}).each {|dc|
                            @doc_list << Document.find(dc.document_id)
    }
  end
  @category_display_name = cname || ''
  session['search_text'] = @category_display_name # required when using the
                                                  # home controller fragment
```

```
                                     # _results.html.erb
   end
```

Notice that I store the selected category name in the user's session in order to make it available to the Home page's _results.html.erb fragment that I reuse on the Categories page. Now is a good time to look at some of the view implementation for the Categories page. The first snippet from the category view handles auto-completion for the category-name input field and uses the custom auto-complete controller method shown in the previous code snippet:

```
<div valign="top">
   <form action="/categories" method="get">
      <%= text_field_with_auto_complete :document_category, :category_name,
                                        :size => 36,
                                        :value => @category_display_name %>
      <input type="submit" value="Browse category"/>
   </form>
</div>
```

The next snippet from the category view processes the documents in the @doc_list array (set by the custom auto-complete controller method) by using the Home page fragment _results.html.erb:

```
<h3>Results:</h3>
<% @doc_list.each {|doc| %>
   <!-- NOTE: I am reusing a method in the home controller and a home
              view fragment 'result_detail': -->
   <%= link_to_remote "#{doc.original_source_uri}",
                               :url => "/home/result_detail?doc_id=#{doc.id}",
                               :update => 'cat_results' %>
   <br/>
<% } %>
<p id="cat_results"></p>
```

You have already seen the implementation for the application's Admin page in Figure 15-3. The implementation of the admin controller method index runs external processes using Ruby's backtick operators:

```
  def index
    @admin_status = ''
    @sphinx_start = nil
    @sphinx_stop = nil
```

I make a Sphinx search query to see if the server is up and running:

```
  begin
    Document.search('hipporandonous')
    @sphinx_stop = true
  rescue
    @sphinx_start = true
  end
```

I check the value of params['command'] to determine which external process to run:

```
@admin_status =
                `rake thinking_sphinx:index` if params['command'] == 'sphinx_index'
if params['command'] == 'sphinx_start'
  @admin_status = `rake thinking_sphinx:start`
  @sphinx_start = nil
  @sphinx_stop = true
end
if params['command'] == 'sphinx_stop'
  @admin_status = `rake thinking_sphinx:stop`
  @sphinx_start = true
  @sphinx_stop = nil
end
@admin_status =
            `script/find_similar_things.rb` if params['command'] == ➡
'find_similar_stuff'
  end
```

You have already seen the implementation of script/find_similar_things.rb. The Sphinx rake tasks were added by the Thinking Sphinx plugin.

The application's Web Service Info page is trivial because it contains only static content, so I will not discuss it. I implement REST-style web services in the next section.

Web-Service APIs Defined in the Web-Service Controller

I think that most Web 3.0 applications should provide web services to access data. For this example, I chose JSON as the format for web-service call results. Returning JSON data is particularly easy in Rails applications because it is a supported rendering format. For web applications that return RDF data, for example, you would have to do additional work: choose or create an appropriate ontology and generate RDF data in application-specific code. I think that RDF data using a well-known ontology has much greater value than JSON, but JSON is fine for this example application. There is little chance of software agents "figuring out" the semantics of JSON data—client programs must be manually written. I implement a SPARQL query endpoint in the next section using D2R for use by Semantic Web–aware clients.

There are three web-service APIs defined in the web-service controller, which perform the following actions: search for documents and return IDs for matching documents, fetch document details from a specified document ID, and return a list of all categories used in the system.

Before showing you the web-service controller code, I will first show you an example call and subsequent results for each of these APIs. For the first example, a search for "java jsp" like

```
http://localhost:3000/web_services/search?q=java+jsp
```

produces this JSON payload:

```
{"doc_ids": [2, 3]}
```

As an example of using the second API, I request detailed information by document ID (id=2, in this example):

```
http://localhost:3000/web_services/interesting_thing_by_id?id=2
```

The preceding request produces this output (shortened for display here):

```
{"id": 2,
 "original_source_uri": "http://knowledgebooks.com",
 "uri": "db/document_repository/d-4712-01-01-3456.html",
 "summary": "My Java JSP based KnowledgeBooks.",
 "plain_text": " ... ",
 "categories": ["computers_ai", "computers_ai_datamining", "computers_ai_nlp",
                "computers_ai_textmining", "computers_programming_java"],
 "similar_things": [[3, 0.126807]]}
```

As an example of the third web-service API, you can get a list of all categories currently in the system using this:

```
http://localhost:3000/web_services/category_names
```

which returns the following JSON data:

```
{"category_names":
    ["computers", "computers_ai", "computers_ai_datamining", ➡
"computers_ai_learning",
     "computers_ai_nlp", "computers_ai_search", "computers_ai_textmining",
     "computers_microsoft", "computers_programming",
     "computers_programming_c++", "computers_programming_java",
     "computers_programming_lisp", "computers_programming_ruby", "health",
     "health_exercise", "health_nutrition", "news", "news_economy", "news_politics",
     "news_war", "news_weather", "religion", "religion_buddhism", ➡
"religion_christianity",
     "religion_hinduism", "religion_islam", "religion_judaism"]}
```

The following listing shows the implementation of the web-service controller, with explanations between the code snippets:

```
class WebServicesController < ApplicationController
```

Implementing the web-service API for returning a list of category names (the third API) uses the class method Document.get_all_category_names and Rails's built-in capabilities for converting Ruby data to JSON and rendering it back to a web-service client:

```
  def category_names
    render :json => { :category_names => Document.get_all_category_names }
  end
```

To return all data for a specific document using the second API defined in the web-service controller, I create an empty hash table and copy in both the document attributes and the data from linked classes:

```ruby
def interesting_thing_by_id
  hash = {}
  id = params[:id]
  if id && doc = Document.find(id.to_i)
    hash = {:id => doc.id,
            :original_source_uri => doc.original_source_uri,
            :uri => doc.uri,
            :summary => doc.summary,
            :plain_text => doc.plain_text,
            :categories => doc.document_categories.collect {|dc| dc.category_name},
            :similar_things => doc.get_similar_document_ids
           }
  end
  render :json => hash
end
```

The search web-service call (the first API) uses the method search that the Thinking Sphinx plugin provides my Document class. For the returned results, I collect the IDs of matching documents:

```ruby
def search
  hash = {}
  if params[:q] && results = Document.search(params[:q])
    hash = {:doc_ids => results.collect {|doc| doc.id}}
  end
  render :json => hash
end
end
```

There are two implementation details that I skipped entirely because they were not the point of this example. However, I want to provide you with some useful references for free plugins and tools that I use:

- *Managing users, including e-mail and SMS new-account authentication*: I sometimes start new Rails projects using Ariejan de Vroom's BaseApp template application (http://github.com/ariejan/baseapp/tree/master). Rails 2.3 provides the new ability to use customizable templates that solve specific development tasks. These templates, written by many developers, offer a wide range of functionality and let you select just what you need.

- *Paginating search results*: I suggest using Mislav Marohni 's pagination plugin that you can download from http://github.com/mislav/will_paginate/tree/master.

We are not finished implementing this example. We don't want a Web 3.0 application that does not support the Semantic Web and Linked Data. I add a SPARQL endpoint in the next section.

SPARQL Endpoint for the Interesting Things Application

I have been making the point in this book that Web 3.0 applications should support Semantic Web technologies. My first thought for making this example "Web 3.0–compliant" was to add a web-service call that would simply dump information about documents and similarity links between documents to RDF as the returned payload. This would be simple to implement, but I have two problems with this approach. First, RDF data that is exported to another system will soon be out of date. Second, I think that the spirit of Web 3.0 dictates that we don't reinvent or recode the same functionality that is already available in other systems or tools.

In Chapter 11 I covered the Linked Data tool D2R, which wraps a relational database as a SPARQL endpoint. I refer you back to that chapter for details on using D2R, but you can get started using only two command-line statements. You need to run the first statement once to create a D2R database mapping file, and you can run the second statement whenever you need SPARQL endpoint services for the Interesting Things web application:

```
generate-mapping -o interesting_things_d2r_mapping.n3              ➥
                        -d com.mysql.jdbc.Driver                   ➥
                        -u root                                    ➥
                        jdbc:mysql://localhost/interesting_things_development
d2r-server interesting_things_d2r_mapping.n3
```

Figure 15-11 shows the D2R SPARQL query browser processing a SPARQL query to return the document IDs and original data source for similar documents.

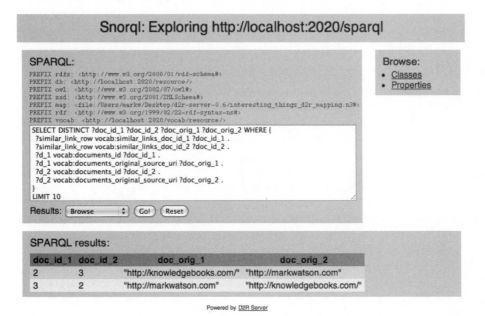

Figure 15-11. *Snorql, the D2R SPARQL query browser*

You can experiment with the D2R SPARQL query browser on your own, but here is another sample SPARQL query to get you started. Here, I want similar original document URIs (or local file paths) and the similarity "strength" between them:

```
SELECT DISTINCT ?doc_orig_1 ?doc_orig_2  ?strength WHERE {
  ?similar_link_row vocab:similar_links_doc_id_1 ?doc_id_1 .
  ?similar_link_row vocab:similar_links_doc_id_2 ?doc_id_2 .
  ?d_1 vocab:documents_id ?doc_id_1 .
  ?d_1 vocab:documents_original_source_uri ?doc_orig_1 .
  ?similar_link_row vocab:similar_links_strength ?strength .
  ?d_2 vocab:documents_id ?doc_id_2 .
  ?d_2 vocab:documents_original_source_uri ?doc_orig_2 .
}
```

This query, with the three test documents in the database as seen in previous examples, produces the following SPARQL query results:

doc_orig_1	doc_orig_2	strength
"http://knowledgebooks.com/"	"http://markwatson.com"	0.126807
"http://markwatson.com"	"http://knowledgebooks.com/"	0.126807

A disadvantage of using D2R instead of writing a custom SPARQL interface is that another process (that is, the D2R service) needs to run. The advantages of using D2R are that you get a lot of functionality with no extra code to write, and you get scalability by separating the SPARQL endpoint functionality out to a different process. In the next section, I discuss how you might scale this Interesting Things application with the SPARQL endpoint to handle a very large number of users.

Scaling Up

If you should happen to write a multiuser version of a system like Interesting Things, you'd want a clear plan for scaling the application to handle many users. The first component of this example web application that you'd need to scale is the database. I discussed scaling MySQL using either master/slave servers or sharding in Chapter 12. If you have relatively few new interesting things added to the system compared to the number of read accesses to the database, then I think using a master/slave architecture makes sense. If you need to also handle a high volume of database inserts, then sharding different tables to different servers is a good plan. Fortunately, for this example web application at least, there are no query joins between tables so sharding should be efficient. A disadvantage of sharding is that it might take extra work getting D2R working with a sharded distributed database.

Scaling up the Rails web application is more straightforward than handling a distributed database. There are several options, but my preferred deployment strategy is to use nginx as a front end to multiple Mongrels. You can split up the Mongrels to run on multiple servers. Using memcached (see Chapter 12) is also recommended. If required, you can also run multiple copies of D2R on different servers.

I hope that the implementation of this example web application has introduced you to new techniques, provided you with ideas for your own projects, and encouraged you to have some fun.

Wrapup

This book is primarily about using Ruby scripts, applications, and web frameworks to write Web 3.0 applications. It seems fitting to end this book with two Ruby on Rails examples. After spending more than ten years developing web applications in Java (as well as less commonly used languages like Common Lisp, Scheme, and Prolog), I find that Rails hits a good compromise for most of my own development. I still view the Java platform to be very good for writing large-scale web applications, but my first choice is Rails because it is so much faster and less expensive to use as a development platform.

It is May 2009 as I write this. The topics that I have covered in this book are my best guess as to what technologies will be important to both you and me in the next few years, as we develop novel new web applications, new systems that are small and focused around specific applications, and new web sites that support use by both human and software agents.

For me, writing is a great pleasure. I also enjoy correspondence with my readers. My e-mail address is markw@markwatson.com. I try to respond to all reader e-mail, but if I am very busy it might take a few days, so please be patient. I maintain the web page http://markwatson.com/books/web3_book/ that provides a link to resources on the Apress web site. I post my replies when I answer questions from readers, so please check the site for questions similar to yours that I might have already answered.

PART 5

■ ■ ■

Appendixes

In Appendix A, you'll get links to Amazon Web Services documentation and learn how to access the Amazon Machine Image (AMI)) that contains all the examples in this book. In Appendix B, you'll see how to publish HTML or RDF based on HTTP request headers. Finally, Appendix C will provide an introduction to RDFa, a notation for embedding RDF in XHTML web pages.

APPENDIX A

■ ■ ■

Using the AMI with Book Examples

I have prepared an Amazon Machine Image (AMI)) with all of the examples in this book. Because I will be periodically updating the AMI, you should search for the latest version. This is simple to do: after you log in to the Amazon Web Services (AWS) Management Console, select "Start an AMI," then choose the "Community AMIs" tab and enter **markbookimage** in the AMI ID search field. Choose the AMI with the largest index. (I'll provide more details shortly.) For updates, you can check both the Apress web site for this book and the errata page for this book on my web site (http://markwatson.com/books/web3_book/).

I assume that you worked through the last part of Chapter 14, so you are familiar with using the s3cmd command-line tools and the AWS Management Console. You can sign up for an AWS account at http://aws.amazon.com/. At the time that I am writing this Appendix, the cost of using an Amazon EC2 small server instance is about $0.10 per hour: very inexpensive, but remember to release the EC2 server instance when you are done with it. If you do not release the instance, you will continually be charged the hourly fee. You can use the AWS Management Console application, as shown in Figure A-1, to double-check that you have stopped all your instances at the end of a work session.

You will need to use the AWS Management Console to create authentication files and set up your laptop to use the Amazon web services. You can find directions in the Amazon EC2 Getting Started Guide (http://docs.amazonwebservices.com/AWSEC2/latest/GettingStartedGuide/). This guide covers prerequisites for setting up your Mac, Linux, or Windows PC system. This setup process initially took me about 15 minutes, but you only need to do it once.

Although it's not necessary for using my AMI, you might also want to bookmark the EC2 Developer Guide (http://docs.amazonwebservices.com/AWSEC2/latest/DeveloperGuide/). If you want to clone my AMI and customize it for your own use (recommended!), I refer you to Amazon's online documentation. The Developer Guide for EC2 is both complete and well-written—recommended reading!

Now I'll explain how to use my AMI for the first time. Once you've created an account for AWS, log in to the AWS Management Console at https://console.aws.amazon.com/. You will need to choose the "Key Pairs" menu item on the bottom-left portion of the screen and generate a key-pair file. Save this file on your laptop; you will need it later to SSH into a running EC2 server instance. You only need to perform this step once.

Navigate to the EC2 Dashboard web page and click "Launch Instance." Then click the "Community AMIs" tab, enter **markbookimage** in the only editable text field, and click the result with the largest index (there might be only one matching AMI if I remove older AMIs after testing the newest image). Start the AMI after specifying that you want to run one instance and that you will be using your gsg-keypair. Specifying the gsg-keypair option will allow you to use SSH to log in to your running instance. After starting an EC2 server instance with an AMI, you can click the link for viewing your running instances. You will see a display like that in Figure A-1.

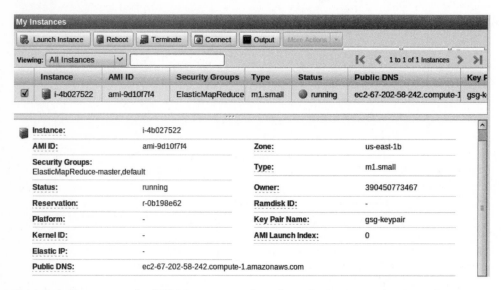

Figure A-1. *You can use the AWS Management Console to view your running instances.*

Notice the public DNS URL in Figure A-1. The first thing that you want to do is SSH into your running EC2 instance; with the URL in Figure A-1 (your public URL will be different), the command would be:

```
ssh -i ~/.ssh/id_rsa-gsg-keypair root@ec2-67-202-58-242.compute-1.amazonaws.com
```

I saved the key-pair file in my local ~/.ssh directory and specified the file location as a command-line option. When you log in as root (no password is required), you will notice a number of README files in the root account's home directory, such as README_tomcat_sesame. These files describe how to run the following examples, which I installed on the AMI:

- Tomcat + Sesame (with preloaded "interesting" RDF repositories)

- Tomcat + Solr (search examples)

- Tomcat + Nutch (search portal examples)

- Franz AllegroGraph (with example Ruby client programs)

- Ruby + Sinatra SPARQL endpoint examples from Chapter 7

- Example Rails portal from the second half of Chapter 15

- Sphinx + MySQL indexing and search (search examples)

- PostgreSQL with native indexing and search (search examples)

- Most command-line example programs in the book

If you view the public DNS URL in a web browser, you will want to use the port numbers mentioned in the README file for each example.

Because it only takes a couple of minutes to select an AMI and start up an EC2 server instance, you can easily experiment with the book examples whenever you want. S3 storage is used to store AMIs, and the cost is fairly low so you can inexpensively clone my AMI and save it as your own. You will then specify your own AMI ID when starting an EC2 server instance.

Warning You need to remember to shut down your EC2 instance when you are done with it. Keeping an instance running continuously costs about $2.40 a day, so shut it down if you are not using it. If you just experiment with the book examples a few times (before deciding which ones to install permanently on your laptop), then your costs should be less than a dollar.

APPENDIX B

■■■

Publishing HTML or RDF Based on HTTP Request Headers

This Appendix provides you with some background on using HTTP headers to determine the information and desired format that web clients are requesting. A good use case for this is a web application that publishes data as HTML pages but returns RDF triples if the HTTP request headers specify that the client needs RDF instead of HTML. I am going to review some details of the Hypertext Transfer Protocol (HTTP)) that will be generally useful.

A key part of the web platform is HTTP. HTTP requests consist of a request line, headers, a blank line to mark the end of the headers, and optional request data. The methods used in request lines are GET, POST, PUT, and DELETE. We have been using the REST architectural style in this book; REST assumes specific meanings for these methods, as shown in Table B-1.

Table B-1. *Mapping of HTTP Methods to Database CRUD Equivalents*

HTTP Method	Database Equivalent
GET	Read
POST	Create
PUT	Either Update or Create
DELETE	Delete

Services respond differently to HTTP requests based on both the method and headers. HTTP requests can use an Accept request header to indicate what type(s) of data they want from a request. Table B-2 shows examples of commonly used types. There are many types like audio, application-specific file formats, and so on that are not required for this book. You can find complete coverage of HTTP header fields at http://www.w3.org/Protocols/rfc2616/rfc2616-sec14.html.

Table B-2. *Examples of HTTP Accept Request Headers*

Accept Request Header	Description
Accept: text/plain, */*	Prefer plain text but can accept any type
Accept: text/html, text/plain;q=0.5	Prefer HTML but can accept plain text
Accept: text/rdf+n3	Request RDF data in N3 format
Accept: text/rdf+xml	Request RDF data in XML format

Depending on the HTTP headers from client requests, you sometimes need to return different types of data—HTML for human readers or RDF for software agents, for example. I will show you how to implement this functionality in Rails web applications.

Returning HTML or RDF Data Depending on HTTP Headers

The code to check headers and detect the type of data requested in an HTTP request header is reasonably easy to write. I will show you an implementation in a Rails web application. A simple way to test services is to use either the wget or curl command-line utility.

You can use the wget utility to perform HTTP GET requests with specific request headers. Here are examples of requesting RDF and HTML:

```
wget --header='accept: text/rdf+n3' http://localhost:8080/
wget --header='accept: text/html' http://localhost:8080/
```

Using curl, you could make the same two requests like this:

```
curl --header 'accept: text/rdf+n3' http://localhost:8080/
curl --header 'accept: text/html' http://localhost:8080/
```

By default, wget writes fetched data to a local file and curl writes fetched data to standard output. Use the -o option with curl to write fetched data to a local file.

Handling Data Requests in a Rails Example

This section contains a simple Rails web application that uses the database table news_articles that we also used in Chapter 8. The web application returns HTML, RDF N3, or RDF XML, depending on the HTTP request headers.

I generate a new project and create a controller index and a view index that I will use to demonstrate handling of custom MIME types:

```
rails rails_http_headers
cd rails_http_headers/
script/generate controller index index
```

Rails has built-in support for handling different request types. Here is example controller code to handle HTML and XML request types:

```
respond_to do |format|
  format.html
  format.xml { render :xml => @news_article.to_xml }
end
```

The `responds_to` method uses MIME types defined in `app/config/initializers/mime_types.rb`. In the preceding code snippet, there is no block passed to `format.html`, so the controller continues to process the request in the usual way: initialize any variables in the controller and call the compiled view template code. Because `format.xml` is passed a code block, that code block is executed and no compiled view code gets called. This all works easily when you're using standard MIME types, but Rails does not define MIME types for RDF so you need to add the following two statements to `app/config/initializers/mime_types.rb`:

```
Mime::Type.register "text/rdf+xml", :rdf
Mime::Type.register "text/rdf+n3", :n3
```

The following code snippet shows the `index` controller using these registered MIME types:

```
class IndexController < ApplicationController
  def index
    respond_to do |format|
      format.html { @request_type = request.format.to_s }
      format.n3   { render :text => "RDF N3 text" }
      format.rdf  { render :text => "RDF XML text" }
    end
  end
end
```

If the request is for HTML data, I set the variable `@request_type` to the requested data format so that it can be shown in the `index` view:

```
<h1>HTML Output</h1>
<p>Request type is: <%= @request_type %></p>
```

The `IndexController` class demonstrates the use of custom MIME types. For RDF N3 or RDF XML format, we bypass any views and directly render text back to the client.

We will now write an ActiveRecord model subclass for the class `News`. I am going to reuse the MySQL database `test` that I used in Chapter 9, specifically the table `news_articles`:

```
mysql> select * from news limit 1;
+---+-----------------+--------------------------------------------+
| id | title          | contents                                   |
+------+---------------+--------------------------------------------+
|    1 | Home Farming News | Both government officials and organic... |
+---+-----------------+--------------------------------------------+
1 row in set (0.42 sec)
```

Without showing all of the details, I am going to create a controller called show for showing individual news stories and generate a model for the News class:

```
script/generate controller show index
script/generate model news
```

I need to edit the model file News.rb to override the default table name and add the methods to_rdf_n3 and to_rdf_xml. In this example, I hard-coded the subject URIs to reference the server localhost, but you would want to change this to the domain name assigned to your server if you deploy this example:

```
class News < ActiveRecord::Base
  set_table_name :news
  def to_rdf_n3
    "@prefix kb: <http://knowledgebooks.com/test#> .\n\n" +
    "<http://localhost:3000/show/index/#{self.id}> kb:title \"#{self.title}\";\n" +
    "                                   kb:contents \"#{self.contents}\".\n"
  end
  def to_rdf_xml
    "<?xml version=\"1.0\" encoding=\"utf-8\"?>
    <rdf:RDF xmlns:kb=\"http://knowledgebooks.com/test#\"
            xmlns:rdf=\"http://www.w3.org/1999/02/22-rdf-syntax-ns#\">
      <rdf:Description rdf:about=\"http://localhost:3000/show/index/#{self.id}\">
        <kb:title>#{self.title}</kb:title>
      </rdf:Description>
      <rdf:Description rdf:about=\"http://localhost:3000/show/index/#{self.id}\">
        <kb:contents>#{self.contents}</kb:contents>
      </rdf:Description>
    </rdf:RDF>"
  end
end
```

I generated a new class ShowController, and its index method processes HTML requests by passing the value of the variable @news to the HTML template app/views/show/index.html. erb. For RDF N3 and RDF XML requests, rendered text is returned to the client through the News class methods to_rdf_n3 and to_rdf_xml:

```
class ShowController < ApplicationController
  def index
    @news = News.find(request[:id])
    respond_to do |format|
      format.html
      format.n3   { render :text => @news.to_rdf_n3 }
      format.rdf  { render :text => @news.to_rdf_xml }
    end
  end
end
```

I added a statement to config/routes.rb so news-article request URLs would be in REST style:

```
map.connect 'show/:id', :controller => 'show', :action => 'index'
```

This allows URLs like http://localhost:3000/show/1 instead of also requiring the controller method name index: http://localhost:3000/show/index/1. Here are two sample client requests with the returned RDF data:

```
$ curl --header 'accept: text/rdf+n3' http://localhost:3000/show/1
@prefix kb: <http://knowledgebooks.com/test#> .

<http://localhost:3000/show/index/1> kb:title "Home Farming News";
                                      kb:contents "Both ➥
government officials and organic food activists agree that promoting home and ➥
community gardens is a first line of defense during national emergencies".

$ curl --header 'accept: text/rdf+xml' http://localhost:3000/show/1
<?xml version="1.0" encoding="utf-8"?>
    <rdf:RDF xmlns:kb="http://knowledgebooks.com/test#"
             xmlns:rdf="http://www.w3.org/1999/02/22-rdf-syntax-ns#">
      <rdf:Description rdf:about="http://localhost:3000/show/index/1">
        <kb:title>Home Farming News</kb:title>
      </rdf:Description>
      <rdf:Description rdf:about="http://localhost:3000/show/index/1">
        <kb:contents>Both government officials and organic food activists agree ➥
that promoting home and community gardens is a first line of defense during ➥
national emergencies</kb:contents>
      </rdf:Description>
    </rdf:RDF>
```

If you access the same URL http://localhost:3000/show/1 in a web browser, you see the output shown in Figure B-1.

Show RDF Data as HTML

1 : Home Farming News

Both government officials and organic food activists agree that promoting home and community gardens is a first line of defense during national emergencies

Figure B-1. *Showing RDF data as HTML*

It is generally good practice to check HTTP request headers to determine the requested data format. Another possible way to determine the requested data format is to end the URL with ".pdf," ".n3," ".xml," and so on. Using a parameter like `?format=xml` is poor style.

I believe that Web 3.0 applications are defined as applications for both human readers and software agents. Because it is easy to add REST-style web APIs and support different returned data formats, consider adding this support to all web applications that you write.

APPENDIX C

■■■

Introducing RDFa

A major theme in this book has been using RDF as a universal data format for use by web and client applications. Another major theme is writing Web 3.0 applications to be useful for both human readers and software agents. Although I don't yet use it in my work, there is a developing standard called RDFa that is a notation for embedding RDF in XHTML web pages. While I think it's a poor idea for people to add RDFa data into the web pages that they manually create, many web pages are generated automatically from structured data sources, and automatically generating RDFa when HTML is generated might prove to be useful. I think that there is a good opportunity to increase the amount of available RDF data by augmenting automatically generated web pages with embedded RDFa. The next version of Drupal (version 7) is expected to have built-in RDFa support. I think it is likely that other content management systems (CMSs) like Plone will get RDFa support in the future. Both Google and Yahoo! also support the use of RDFa.

One large RDFa data deployment is the Riese project (http://riese.joanneum.at/). The clever slogan on its web page, "Humans & machines welcome!", demonstrates the spirit of Web 3.0. Riese is part of the EuroStat Data Set Effort that aims to make public data easily available.

You can find a complete tutorial on RDFa at http://www.w3.org/TR/xhtml-rdfa-primer/.

RDFa is a set of attributes that you can add, along with values, to XHTML element tags. Here is a short example:

```
<div xmlns:dc="http://purl.org/dc/elements/1.1">
  <h1 property="dc:title">
      <a href="http://markwatson.com/opensource/" rel="cite">
         Open Source Projects
      </a>
  </h1>
  <h3 property="dc:creator">Mark Watson</h3>
<p>All content on this site is licensed under either the
    <a rel="license" href="http://www.gnu.org/copyleft/gpl.html">
       GPL License
    </a>
    or the
    <a rel="license" href="http://www.gnu.org/copyleft/lgpl.html">
       LGPL License
    </a>
```

```
    </p>s
</div>
```

In this example, I used properties and types defined in the Dublin Core Metadata Initiative project, with the dc: namespace prefix. You generally need to qualify properties and types with namespace prefixes. But the properties cite and license, which refer to citation links and license links, don't require namespace prefixes because they're in the default RDFa namespace. Although RDFa is not syntactically correct in HTML 4 (because HTML 4 is not extensible), on a practical level, RDFa can be used with HTML.

The RDFa Ruby Gem

The RDFa example in the preceding section used two vocabularies: a built-in vocabulary that requires no namespace prefixes, and the Dublin Core namespace. The RDFa gem, written by Cédric Mesnage, defines utility methods that can be used in Rails controllers and view templates for generating XHTML with embedded RDFa.

Table C-1 (from the RDFa gem documentation) lists a few of the RDFa gem utility methods that generate XHTML for common RDF classes.

Table C-1. *RDFa Gem Utility Methods for RDF Classes*

Class with Prefix	RDFa Method
foaf:Person	rdfa_person
rdfs:Resource	rdfa_resource
sioc:Community	rdfa_community
rdf:List	rdfa_list
sioc:Forum	rdfa_forum
sioc:Post	rdfa_post

You will see some examples using RDFa gem utility methods in a sample Rails application in the next section. Table C-2 (again, from the RDFa documentation) lists a few of the utility methods to generate XHTML for commonly used RDF properties.

Table C-2. *RDFa Gem Utility Methods for RDF Properties*

Property with Prefix	RDFa Method
dcel:source	rdfa_source
foaf:knows	rdfa_knows
rdfs:label	rdfa_label
foaf:homepage	rdfa_homepage
foaf:weblog	rdfa_weblog
foaf:name	rdfa_name
rdfs:seealso	rdfa_seealso

Property with Prefix	RDFa Method
dcel:title	rdfa_title
sioc:content	rdfa_content
dcel:description	rdfa_description
dcel:creator	rdfa_creator
foaf:mbox	rdfa_mbox
dcel:date	rdfa_date
dcterms:license	rdfa_license
rdfs:comment	rdfa_comment

You can find a complete list of the vocabulary supported by the RDFa gem at http://rdfa.rubyforge.org/vocabularies.html. I implement a simple Rails application in the next section that uses the RDFa gem.

Implementing a Rails Application Using the RDFa Gem

I wrote an example Rails application that uses the RDFa gem, which you need to install with gem install rdfa. The source code for this example is in src/appendices/C/rails_rdfa (you can find the code samples for this Appendix in the Source Code/Download area of the Apress web site). I generated an index controller.

I made no changes to the controller. I edited the default view for the index controller to use the helper methods defined in this gem. I have been using the prefix dc: in this book for the Dublin Core Metadata; the RDFa gem uses dcel:: as the prefix. The edited view file looks like this:

```
<%= rdfa_html %>

<%= rdfa_name "Mark Watson" %>     <br />

<% rdfa_class 'rdfs:Resource' do %>
  <a href="http://cookingspace.com">Food nutrients</a>
<%end%>   <br />

<% rdfa_triple 'dcel:title', 'Healthy Recipes', "http://cookingspace.com" do %>
  Look up recipes with their nutrients
<%end%>   <br />

<%= rdfa_link_to 'dcel:title', "Mark Watson: Ruby and Java Consultant",
                     'http://markwatson.com' %>
```

The method rdfa_html generates XML namespace definitions. The generated HTML is shown here:

```
<html xmlns:dcel="http://purl.org/dc/elements/1.1/"
         xmlns:dcterms="http://purl.org/dc/terms/"
         xmlns:dctypes="http://purl.org/dc/dcmitype/"
         xmlns:doap="http://usefulinc.com/ns/doap#"
         xmlns:foaf="http://xmlns.com/foaf/0.1/"
         xmlns:rdf="http://www.w3.org/1999/02/22-rdf-syntax-ns#"
         xmlns:rdfs="http://www.w3.org/2000/01/rdf-schema#"
         xmlns:sioc="http://rdfs.org/sioc/ns#"
         xmlns:swrc="http://swrc.ontoware.org/ontology#"
         xmlns="http://www.w3.org/1999/xhtml">

<span property="foaf:name">Mark Watson</span>  <br />

<span class="rdfs:Resource" rel="rdf:li">
  <a href="http://cookingspace.com">Food nutrients</a>
</span>

<br />

<span about="http://cookingspace.com" content="Healthy Recipes"
         property="dcel:title">Look up recipes with their nutrients</span>

<br />
<a href="http://markwatson.com" rel="dcel:title">
   Mark Watson: Ruby and Java Consultant
</a>
```

Figure C-1 shows the generated web page for this view.

Mark Watson
Food nutrients
Look up recipes with their nutrients
Mark Watson: Ruby and Java Consultant

Figure C-1. *Web page showing the default view of the index controller*

My use of Cédric Mesnage's gem is simple because I just added static RDFa (using the gem utility methods) to a view. For a more realistic example, you could use a publishing database to create an HTML table containing book and author information that has embedded RDFa data. I suggest that you look at the Riese project (http://riese.joanneum.at/) that I mentioned previously as a good example of a system using RDFa. You can "View Source" on your web browser to see the embedded RDFa when viewing economic data on the Riese site.

As I mentioned, I have not yet used RDFa in any projects, but I like the idea of using it to create dual-purpose web applications. I think wide-scale adoption of RDFa will depend on its implementation in authoring tools like Drupal and Plone as well as web frameworks like Rails, Django, and so on.

Index

You Need the Companion eBook

Your purchase of this book entitles you to buy the companion PDF-version eBook for only $10. Take the weightless companion with you anywhere.

We believe this Apress title will prove so indispensable that you'll want to carry it with you everywhere, which is why we are offering the companion eBook (in PDF format) for $10 to customers who purchase this book now. Convenient and fully searchable, the PDF version of any content-rich, page-heavy Apress book makes a valuable addition to your programming library. You can easily find and copy code—or perform examples by quickly toggling between instructions and the application. Even simultaneously tackling a donut, diet soda, and complex code becomes simplified with hands-free eBooks!

Once you purchase your book, getting the $10 companion eBook is simple:

❶ Visit **www.apress.com/promo/tendollars/**.

❷ Complete a basic registration form to receive a randomly generated question about this title.

❸ Answer the question correctly in 60 seconds, and you will receive a promotional code to redeem for the $10.00 eBook.

THE EXPERT'S VOICE™

2855 TELEGRAPH AVENUE | SUITE 600 | BERKELEY, CA 94705

Offer valid through 1/10.